T0230919

LOW-LEVEL RADIATION and IMMUNE SYSTEM DAMAGE

An Atomic Era Legacy

LOW-LEVEL RADIATION and IMMUNE SYSTEM DAMAGE

An Atomic Era Legacy

Joseph J. Mangano

The Radiation and Public Health Project
New York, New York

CRC Press
Taylor & Francis Group
Boca Raton London New York

CRC Press is an imprint of the
Taylor & Francis Group, an **informa** business

First published 1999 by CRC Press
Taylor & Francis Group
6000 Broken Sound Parkway NW, Suite 300
Boca Raton, FL 33487-2742

Reissued 2018 by CRC Press

Library of Congress Cataloging-in-Publication Data

Mangano, Joseph J., 1956 –
    Low-level radiation and immune system damage : an atomic era legacy / Joseph J. Mangano.
        p.   cm.
    Includes bibliographical references and index.
    ISBN 1-56670-334-4 (alk. paper)
    1. Low-level radiation—Health aspects—United States.   2. Immunosuppression—United States.   I. Title.
    RA569.M36 1998
    616.9'897—dc21   98-18138

A Library of Congress record exists under LC control number: 98018138

ISBN 13: 978-1-315-89510-9 (hbk)
ISBN 13: 978-1-351-07420-9 (ebk)

Visit the Taylor & Francis Web site at http://www.taylorandfrancis.com and the
CRC Press Web site at http://www.crcpress.com

The Author

Photo by Mary Lynn Melton

Joseph Mangano is a public health administrator and researcher who has been a consultant with the Radiation and Public Health Project, a non-profit radiation research group, since 1989. He has published a number of articles in medical journals, including the renowned *Lancet* and *British Medical Journal*, on health effects of low-level radiation exposure. His work has been presented at four international conferences on radiation. Mr. Mangano's research has shown repeatedly that small doses of radiation from nuclear plants or bomb tests are connected with adverse health effects in humans.

Mr. Mangano received a Master's in Public Health from the University of North Carolina at Chapel Hill, and a Master's in Business Administration from Fordham University in New York. He lives in Brooklyn, New York.

Dedication

To the many fellow Baby Boomers with whom I grew up, went to school, worked, and lived: for those of you who have been stricken with immune illnesses, I hope that society will learn from your unfortunate example and shrink the sick-call rolls in the future; and for those of you who are healthy, I hope you will go on to live long and prosperous lives.

To the generation of beautiful young American children born in the 1980s and 1990s, including four special nephews and a special niece: I hope you and your peers grow strong and hearty, and avoid some of the ill health that burdened earlier generations.

And to PHS, a very special Baby Boomer, my humble thanks for always being a great inspiration.

Contents

Preface

I'm not sure the genesis of a book about the health effects of radiation can be traced to a single point of origin. Maybe you have to go all the way back to 1895, with the discovery of X-rays by Roentgen. Another key time is the World War II period, when the modern nuclear era began with the introduction of weapons of mass destruction. Any of a number of events during this time could well serve as a beginning; the first chain reaction at the University of Chicago in 1942, or the detonation of the first bombs at Alamogordo, Hiroshima, and Nagasaki three years later all make excellent candidates.

Maybe the real start takes place at my birth in the mid-1950s, a time when the United States government was exploding dozens of nuclear devices over the Nevada desert. My entry into the field of public health some 20 years later serves as another possible basis for the book's beginning. It could be argued that my developing a serious immune disorder in the mid-1980s, a time when hundreds of thousands of other young American adults were undergoing the same experience, served as the impetus for preparing this volume. A final nominee is 1989, when curiosity over my chronic fatigue syndrome led me to encounter two dedicated scientists, Jay Gould and Ernest Sternglass, who had been grappling with determining the true extent of damage to human health caused by man-made radioactivity. Ever since then, these two men and I have exchanged numerous ideas and information that are found throughout the book.

So perhaps there is no single origination point. Or maybe each of the events mentioned contributed in its own way to the need for this work and for my idea to produce it. Until 1989, I knew that atomic energy carried potential health hazards, but like millions of Americans I understood few of the details. However, much of my last nine years has been spent trying to better understand such hazards; and what I have learned has endowed me with an ironclad commitment to reduce the hazard so that fewer Americans will suffer.

The book explores how events in the nuclear era affected the health of two groups of people: the Americans generally born between the end of World War II and the mid-1960s (the so-called "Baby Boomers") and those born after the early 1980s. A lot has been written about the collective experiences of particular generations of Americans, and I am certainly not the first to look at the Baby Boom generation. However, I will venture that I may be the first to target the generation born in the past decade or so, mostly because they are still early in their lives and have yet to make their mark on society.

This book (with apologies to Dickens' *A Christmas Carol*) is about the past, present, and future. It examines what has transpired during the nuclear era, covering the evolution of atomic weapons and civilian nuclear power development. It looks at current (as well as past) nuclear policies and health trends that may be a result of these policies. And finally, it speculates about the 21st century, about what kind of America, or even what kind of a world, will emerge as a result of decisions regarding nuclear energy.

This is very much a work rooted in personal concerns. Naturally, as a Baby Boomer myself, I can identify with the suffering of my generation from immune system-related diseases that may be due (at least in part) to radiation exposure. These are the people I went to school with and played with, family and friends with whom I've lived with and worked. And unfortunately, in adulthood, their ranks increasingly include people with AIDS, chronic fatigue syndrome, and cancer. The generation born in the last decade or so is also one I take to heart. These are my nephews and niece, my friends' children, all the beautiful and shining little faces that are emerging among us. They have barely begun to live, and have every right to long, healthy, prosperous lives. I have spent many an hour poring through musty old volumes of health statistics, seeing not just numbers, but the faces of the living healthy, the living sick, and the dead.

I feel quite fortunate to be able to assemble a book like this. As a public health administrator, I have been trained to understand health issues and implement policies that best assure the health of all people. As someone with considerable experience in epidemiology and health information, I can calculate measures of health status and evaluate the effects of health hazards. And as someone with a longstanding interest in current events — my undergraduate studies, ongoing reading, and the book I wrote years ago on important events of 1964 are examples — I can fit the issue of radiation's effects into a historical context.

The information in these pages raises as many questions as it answers. Because nuclear science is still relatively new, because conclusively attributing health effects to a single factor such as radiation is difficult to do, and because the understanding of radiation's true effects has long been clouded by Cold War politics and economics, some of the content will be controversial. Because the book finds that radiation may have had greater adverse health effects on the two generations mentioned than previously believed, a natural reaction may be to doubt or deny the findings. Even free from the shackles of Cold War censorship, intimidation, manipulation, and distortion, it is a natural human impulse to dismiss bad news.

But, as we often learn in life, the truth hurts, and can be tough to swallow. Unwillingness to face up to the unpleasant truth has always had adverse consequences for the human race, not just in the case of radiation, but in many areas. However, with man's knowledge of the atom carrying such enormous destructive potential, the search for the truth must be vigorous and unrelenting, lest we face even more serious consequences than in the past half century. The only path to this truth is an iron-clad commitment to learning, no matter how disturbing the results.

In his famed 1963 *Letter From a Birmingham Jail*, Dr. Martin Luther King declared

> Just as Socrates felt it was necessary to create a tension in the mind so that individuals could rise from the bondage of myths and half-truths to the unfettered realm of creative analysis and objective appraisal, so we must see the need for nonviolent gadflies to create the kind of tension in society that will help men rise from the dark depths of prejudice and racism to the majestic heights of understanding and brotherhood.*

King was speaking about the tension needed to achieve justice in race relations. I submit we need an equally strong tension in people's minds for society to avoid serious health consequences from atomic energy. Without this tension, our health and vitality may be crippled, and our survival may even be threatened.

As a health professional I have learned that diseases don't kill and maim, ignorance does. In the 1300s, millions of Europeans believed they were being struck down with some vague but terrible illness sent to them as punishment by an angry God. Many years later, it was determined that this disease was a bacterial infection called *Pasturella pestis*, or bubonic plague, which was spread to humans by fleas that had bitten infected rats. Subsequently, scientists developed antibiotic drugs to successfully combat the disease. It wasn't just a disease that wiped out one third of the European population in a decade. In retrospect, an accurate diagnosis and an adequate supply of tetracycline or other suitable antibiotic would have reduced the massive suffering to virtually nothing. A dedication to understanding radiation and its effects will reduce the ignorance that continues to hurt people by the thousands.

My generation, the Baby Boomers, are now young adults. The generation born since the early 1980s will soon begin to reach adulthood, and assume their place in society. In all likelihood, the challenge of understanding the nuclear issue and creating a sound nuclear policy will fall largely on the shoulders of the Baby Boomers and younger generations. The events of the past have left a deep nuclear stain on society, a stain that an older generation could not, or would not, reduce or eliminate. However, there still is a chance of averting the suffering of the past and present, the suffering that I attempt to recount in these pages. By writing this, I hope to inspire the leaders of our American society to make sound decisions affecting the future health of Americans. But I also hope to reach all Americans, because as members of a democratic society, we are all leaders and the right to make decisions belongs to all of us.

* Oates, S., *Let the Trumpet Sound*, Mentor Books, New York, 1982, 216.

part one

---

*Political decisions usher
in man-made radiation*

chapter one

Introduction

We often hear that America is in the midst of a health crisis, which usually means costs are too high, many people lack health insurance, or insurance companies are preventing Americans from freely choosing their physicians and hospitals, and not paying for needed services. Aside from the AIDS epidemic, however, few believe there is a crisis in the actual well-being of Americans. The most commonly used statistical measures show that people are in better health than ever. According to the National Center for Health Statistics, death rates in the 1990s are at an all-time low and still dropping, meaning more people than ever are living well into old age. Accordingly, life expectancy continues to grow.

However, in the last several decades, rates of many immune-related diseases have been steadily rising, including cancer, asthma, allergies, septicemia, and hypothyroidism. Better methods of treatment help keep death rates down, but the fact remains that the number of persons developing these conditions has been rising for a long time, and Americans often don't know about these increases or accept them as a "normal" part of modern life. There is certainly no perception of a crisis.

Such a development will certainly fit any definition of crisis. No consistently adverse trends in a group of diseases relying on the immune system's defenses involving millions of people should be ignored, even though explaining the trends is not easily done. It is quite possible that many factors are contributors, including man-made ones, unique to the late 20th century. The University of Chicago's Samuel Epstein believes that about 80% of cancers are the result of man-made factors that do not occur in nature.[1] His book, *The Politics of Cancer*, refers to numerous studies linking environmental hazards to an increased risk of one or more cancers. The culprits include tobacco, polychlorinated biphenyls (PCBs), asbestos, pesticides, benzene, carbon monoxide, chlorofluorocarbons (CFCs), various plastic products, medical and dental X-rays, and an extensive host of others.

This book considers yet another possible factor that first entered the environment in the 1940s: man-made radioactive chemicals produced from atomic bomb tests, nuclear power production, and nuclear waste. The debate over the health effects of radiation has been ongoing since the first atomic bombs were dropped in 1945. While the experience of Hiroshima and Nagasaki convinced experts that *high-level* exposures produced horrifying consequences, effects of *low-level* radiation remain controversial. Until the late 1950s, scientists assumed that low-level radiation exposure was harmless, based on calculations derived from the atomic bomb experience and use of high-dose X-rays. However, beginning in 1956, the field was rocked by several discoveries. British physician Alice Stewart published startling research showing that fetuses exposed to radiation from their mother's (low-level) pelvic X-rays had a much greater chance of developing childhood cancer.[2] After several years of fierce debate, Stewart's findings were duplicated by another study, and the practice of administering pelvic X-rays to pregnant women was stopped. Also in the late 1950s, when dozens of above-ground atomic tests were being conducted in the U.S. and the Soviet Union, two scientists separately stepped forward to proclaim that low-level test fallout, steadily consumed in the diet, was harming the immune system of *all* Americans, Soviets, and other peoples, which would eventually result in large rises in immune disease.[3,4] Although these scientists (Linus Pauling of the U.S., Andrei Sakharov of the Soviet Union) were praised by some and reviled by others, including their respective governments, their message helped to pave the way for the 1963 treaty banning any future atomic bomb tests above the ground.

In 1971, Canadian physician and physicist Abram Petkau made a discovery that formed a theoretical basis for low-level radiation's damage. Petkau found that only a small amount of radioactive sodium was needed to break the membrane of cells, much smaller than previously believed. When radioactive materials break the cell membrane, they can then damage the cell's reproductive function and disrupt the genetic material in the cell nucleus, raising the risk of diseases such as cancer and genetic disorders.[5]

Petkau's discovery came at the same time that a few pioneering scientists began to speak out against nuclear power plants, which were proliferating rapidly. The best known of these were John Gofman of the Lawrence Livermore Laboratory in Berkeley, CA, and Ernest Sternglass of the University of Pittsburgh. They contended that while plant accidents presented great danger to society, the low-level emissions *routinely* discharged from the reactors were causing an ever-growing threat to human health by damaging immune systems. Their beliefs ignited a debate that continues to this day.

Anyone who dared challenge the notion that low-level radiation was safe was and still is sure to meet with strong opposition from two powerful sources. The first is the federal government, specifically the U.S. Department of Energy and its Nuclear Regulatory Commission (formerly the Atomic Energy Commission), which is in charge of regulating and ensuring the safety of nuclear reactors. The second is the nuclear industry, which consists of reactor manufacturers like Westinghouse and General Electric, defense contractors like Union Carbide and Martin Marietta operating nuclear weapons plants, and utilities such as (New York's) Consolidated Edison, Detroit Edison, and (New England's) Northeast Utilities operating nuclear power plants. Citizens groups and independent scientists often find these industries a powerful obstacle to changing national nuclear policy and practices.

Although low-level radiation's consequences are still hotly contested, this book brings considerable new information that figures strongly in the discussion. Chapters 2, 3, and 4 examine the history of the atomic age. Specifically, the political and economic decisions that led to the growth of nuclear power in the U.S., such as the need to end World War II and the Cold War arms race, are reviewed. The chapters also look at the types and amounts of radioactive chemicals introduced into the U.S. environment as a result of these policies. Finally, considerable material is devoted to the evolving debate over health effects of low-level radiation, and how Cold War politics helped suppress a thorough, objective investigation for many years.

Studies of effects of an environmental pollutant such as radiation on health often compare the "dose" vs. the "response." If Chapters 2, 3, and 4 describe the dose, the next five chapters address the response. This material analyzes trends in immune-related disease for the Baby Boom generation, born roughly between 1945 and 1964, compared to earlier and later generations.

Why look at Baby Boomers specifically, especially when everyone living in the atomic age has been exposed to man-made radiation? The greatest exposures occurred from the mid-1940s to the mid-1960s; atomic bombs were mostly tested above ground rather than underground, and nuclear weapons and power plants released the highest levels of radioactive emissions. During these years, the Baby Boomers were fetuses, infants, and young children, when the immature and fast-developing immune system is most susceptible to damage from hazardous sources like radiation. So Baby Boomers, now mostly in their 30s and 40s, suffered more immune damage than did other age groups.

Because of enhanced social and medical conditions, Americans born after World War II should be the healthiest in the history of the world. Many of the terrible infectious diseases of earlier times — polio, diphtheria, Spanish flu, tuberculosis, smallpox, cholera, etc. — have either been eradicated or controlled. A greater supply of food and improved nutritional information has been available to the post-war population. The group didn't have to live through the deprivation of a Great Depression or the carnage of a world war. The technology governing food safety, water quality, and sanitation is better than ever. Housing has improved greatly from the sub-standard conditions — lack of running water, inadequate bathing facilities, sub-standard ventilation — that many had to endure years ago. The population is better educated than ever. Americans are exercising in record numbers. Even a ferocious immune menace such as cigarette smoking has been reduced from 42 to 25 percent of the adult population since the mid-1960s, according to the National Center for Health Statistics. Finally, a massive investment in the health system since World War II — covering diagnosis, treatment, and some prevention — should help make younger Americans the healthiest of all time.

But when it comes to immune diseases, that just isn't the case. Now, more than ever, young American adults are suffering the ravages of cancerous tumors that attempt to crush their immune defenses.

A new disease appeared on the scene in 1981, and has since devastated the immune systems of hundreds of thousands of (mostly young adult) Americans, as part of the world pandemic known as AIDS.

In the mid-1980s, a strange disease began sending thousands of Americans, about 70% of them young adults, staggering to their beds suffering from overwhelming fatigue and a variety of other debilitating symptoms. The disease was initially mocked as "depression," "just stress," "yuppie flu," or "hysteria." Then it was found to be an old disease that had suddenly struck hundreds of thousands of new victims; and later it was correlated with a variety of immunological and hormonal abnormalities. Today, the disease is commonly known as chronic fatigue syndrome, or CFS.

Other abnormalities that may be linked directly or indirectly to the immune response are on the rise among young adults. U.S. government statistics and independent research shows that asthma, other allergies, septicemia, infertility, low sperm counts, and obesity are all increasing, and science has yet to pinpoint an explanation.

These current trends follow a long, spotty history of immune-related disease for the Baby Boomers. According to the National Center for Health Statistics, the percent of babies born below normal weight (under 5½ pounds) actually increased in the 1950s and early 1960s, despite improved technology to enhance fetal and maternal health. No progress was made in fetal deaths, infant deaths, and birth defect-related deaths during these years. These trends are not well known to the general population or even to scientists, but may represent (among other things) an outcome of inadequate immune response on the part of the mother and/or the baby. As children, the pattern of unusually large rises of immune-related disease continued for the Boomers. Cancer, leukemia, septicemia, measles, scarlet fever, and encephalitis were all more prevalent than in previous generations. So the current patterns are not a recent phenomenon, but the continuation of a trend going back several decades.

Much of the data used in this book are national, and since many Americans live far from any nuclear weapons or power plant, some may believe this eliminates radiation as a cause of rising immune disease. However, whenever available, the same information is given for the areas closest to nuclear plants, or those with the heaviest bomb test fallout. Almost always, effects in these exposed populations are the same or larger than the nation as a whole, meaning that exposure to nuclear fission products must be considered as a factor.

While this book is an examination of medical and statistical information describing each of these phenomena, we must not forget that the health and well-being of real people are at stake. The diseases covered in this book generally do not include minor immune problems like the common cold or influenza, but focus on disorders that do serious harm to their victims, including the young. Many of those who are stricken are anonymous, but some are cases of well-known, successful, and otherwise healthy people. For example, famed college basketball coach Jim Valvano became one of the post-war generation to fall victim to the wave of immune disease. In 1992, the 46-year-old Valvano, always an energetic man in excellent health, was diagnosed with metastatic adenocarcinoma, which proved fatal only 10 months later. An article published just weeks before his death described some of the physical and emotional terror that is the mark of cancer. There is that awful moment when the diagnosis is first learned:

> He was still laughing while in the MRI tube last June at Duke University hospital, joking through the intercom with the nurses about the heavy-metal music they were pumping into his headphones as they scanned his spine to see if he had damaged a disk, when the radiologist glanced at the image appearing on the screen, and suddenly the laughter stopped and the nurses fell silent. And the dread, the sick dread began to spread through his stomach as the radiologist quietly said, "Come with me, Coach." And then: "Let me show you a picture of a healthy spine, Coach. Now look at yours."
>
> The vertebrae in his were black where the others were white. And the dread went up in Vee's chest, wrapped around his ribs and his throat, but he squeezed out another joke: "You forgot to use the flash."
>
> No laughter. "Coach, this is just how we see it in the textbook. Coach, I'm 90 percent sure this is cancer."

Valvano's last months were a torturous countdown to his death. The only glimmer of hope lay in chemotherapy, which may help cancer victims but is often as physically punishing as the disease itself:

> ...then he would go to the office of a doctor who tried to be cheerful but who saw 40 cancer patients a day; and then he would be sent to the third floor to lie down again and have Velban, a cell killer, pushed into his veins through the port in the hope that it would kill as many cancer cells as healthy cells. Finally, he would limp out clutching Pam (his wife) for support, his body bent as if beaten with a bat.[6]

Cancers of a variety of organ systems are affecting young and middle-aged adults in greater numbers, but perhaps no cancer has increased more sharply than has breast cancer. Surprisingly, death rates remain relatively unchanged from a generation ago because of more effective treatments; but the number of breast cancer survivors — who have undergone surgery, radiation, chemotherapy, or a combination of these agonizing treatments — is growing fast. New York writer Joyce Wadler, a 43-year-old breast cancer survivor, described her 1991 surgical experience. Wadler underwent a lumpectomy, which is a partial removal of breast tissue, and not nearly as invasive as mastectomy. The surgery was later followed by an extended and debilitating regimen of radiation and chemotherapy:

> Then she suggests something else: a sedative called Versed, which will feel like very strong Valium. Next thing I know, I am waking up in the recovery room, slightly nauseated, drifting in and out of consciousness. There's a slight burning, pulling pain near my left armpit when I try to move, and I feel encumbered by tubing: on my right a pouch of a dextrose-and-saline solution is hanging from a metal pole, feeding a clear solution through a needle-thin IV into a vein on the top of my right hand; coming out of my left side, about four inches under my armpit, there's a peculiarly long rope of plastic tubing attached to a plastic pouch. I am so woozy with morphine I cannot raise my head.[7]

The number of new AIDS cases striking young and middle-aged adults each year is approaching the number of cancer cases. AIDS is a disease that matches the hideous experience of cancer, but unlike cancer, has so far proved to be inevitably fatal within several years of the onset of symptoms, although recently discovered treatments give hope that this will change in the future. The horror of this short period between diagnosis and death has been chronicled many times over. One of the more definitive accounts of AIDS was *And the Band Played On* by Randy Shilts, who himself succumbed to the disease in 1994. Shilts introduced readers to a number of AIDS sufferers, perhaps none as memorable as Bill Kraus, a political activist in San Francisco. In 1984, Kraus, a 37-year-old gay man, was diagnosed with AIDS after noticing a purple spot on his thigh, which proved to be Kaposi's sarcoma. His remaining 15 months of life featured powerful emotional swings; an "I can beat it" attitude shortly after his diagnosis was replaced by "I don't think I'm going to make it" several months after faring poorly with an experimental drug. A description of his final days reflects the brutality of the disease:

> By Christmas, Bill was having a hard time keeping
> food down and suffering from oppressive diarrhea. He
> weighed 120 pounds. Headaches pounded his brain
> like heavy wooden mallets.
>
> (just before he was taken to the hospital before he died)
> Bill was sprawled on the floor where he had fallen.
>
> "I want my glasses!" Bill shouted.
>
> All anyone could hear, however, was "Glubish nein
> ubles sesmag."
>
> It was as if Bill was speaking some strange mix of
> German and gibberish. Somewhere between his brain
> and his mouth, his words were lost.
>
> "Bill, you're not speaking English," Dennis said.
>
> A sheepish grin crossed Bill's lips.
>
> "Gluck eye bub glenish?" he asked tentatively.
>
> "No," Dennis said. "You're not speaking English. We
> can't understand anything you're saying."[8]

No cancer or AIDS case is without enormous anguish. Some of the less
ferocious cancers, such as thyroid and testicular tumors, are usually not
fatal and are not nearly as common as breast, lung, or prostate cancer.
However, these malignancies are becoming more common in young adults,
all of whom must undergo great physical and emotional trauma for the
rest of their lives. One example is Millie Smith, who grew up in Pasco,
WA, downwind of the nearby Hanford nuclear site. In 1986, the year the
U.S. government revealed that massive doses of radioactive iodine were
released into the atmosphere from Hanford in the 1940s, the 39-year-old
Smith received some bad news:

> In 1986 I learned the truth about Hanford, and shortly
> before Christmas of that year my metastasized thyroid
> cancer was discovered. The doctors said I could have
> had it for twenty years. In horrified disbelief I sought
> a second opinion, and that doctor told me that I prob-
> ably would not last for two more years and that im-
> mediate surgery was my only hope.

> Thyroid surgery is very delicate, and there are differences of opinion regarding the best procedure. Possible complications from the surgery include paralysis of the vocal cords and tracheostomy. I did not want to have a tracheostomy. I did not want to lose my thyroid. I did not want to die. I sent for my medical report and learned that my jugular vein was also involved, and in fact was barely working. No one could tell me what that meant, either. Terrified though I was, I finally agreed to have the surgery. Later I learned that my surgery was considered nearly miraculous. The cancer had spread throughout my neck and upper chest, and there were tumors on my laryngeal nerve, which also had to be removed.[9]

Smith was lucky. She survived, joining the growing ranks of middle-aged Americans who have been afflicted with cancer.

Although not fatal like cancer or AIDS, chronic fatigue syndrome has partially or completely disabled several hundred thousand Americans, a number of them educated and skilled contributors to society (which may have initially earned the disease the derisive name "yuppie flu"). CFS usually afflicts those who are in good health, such as this young man stricken during Thanksgiving 1986:

> I was eating breakfast and I felt sick, like the flu came on. I can remember it very distinctly…during that ten-day stay at my in-laws' house, I spent forty to sixty percent of the time in bed trying to sleep it off. I never really got better.

The disease is marked not only by a wide range of symptoms, but by an exasperating series of relapses which usually follow even modest over-exertion, as explained by the same young man:

> If I push myself I can go out maybe one or two nights a week, but I pay for it dearly. The next day I'm done, that's it; I'm sicker. It's a real effort for me to do anything. I'm too tired and too lazy to eat well; I'd rather stay in my chair.[10]

Compounding the physical suffering of CFS is the brutal rejection experienced at the hands of many skeptics in the medical and public health communities who believe that the disease is stress or depression, and dismiss patients who present with the disease. The following story of a young woman with CFS, suffering from the familiar symptoms of fever, swollen lymph nodes, sore throat, joint/muscle pain and weakness, headache, dizziness, and fatigue includes a forgettable visit to an infectious disease specialist:

> The interview lasted twenty minutes, the physical examination ten minutes more. By the end, the specialist did not seem particularly impressed with Alison. He acted disappointed, as if he hoped to find an enormous abscess and it wasn't there. Two weeks later, she learned the tests were normal. He asked a few questions about how she was feeling, but did not seem to hear when she said that she felt as bad as ever. He did a brief physical examination which appeared to be a waste of time, and asked [Alison's husband] Ron if his wife was under any unusual stress. Ron answered that she hadn't been prior to the illness. The specialist suggested that she have a psychiatric interview as fatigue was a prominent symptom in depression. He felt that depression was probably the cause of all the symptoms. Alison and Ron were stunned. They returned home without speaking.[11]

These stories, harrowing as they are, are all about young and middle-aged adults, born in the two decades after World War II. The increasing suffering of this group, throughout infancy, childhood, adolescence, and adulthood, represents a tragedy with enormous repercussions for American society.

The list of victims is a long one that is growing by the day. Many are anonymous, John Q. Public types who are unknown beyond their family and friends. But mixed in with the crowd is a growing number of famous younger Americans afflicted with immune diseases. The oldest of the postwar births include a rising number of immune casualties. Consider the following lists of well-known personalities born between 1946 and 1953 who died of cancer or AIDS between 1991 and 1998 alone:

Cancer
Kathy Ahern, pro golfer, died 1996, age 47
Lee Atwater, political adviser, died 1991, age 40
Mollie Beattie, Clinton wildlife official, died 1996, age 49
Dan Duva, boxing promoter, died 1996, age 44
Tim Gullikson, tennis champion and coach, died 1996, age 44
Rebecca LaBrecque, concert pianist, died 1996, age 45
Nancy LaMott, cabaret singer, died 1995, age 43
Laura Nyro, composer and singer, died 1997, age 49
Esther Rome, author of "Our Bodies, Ourselves" died 1995, age 49
Dawn Steel, first female movie studio head, died 1997, age 51
Brandon Tartikoff, NBC President, died 1997, age 48

Jim Valvano, basketball coach, died 1993, age 47
Carl Wilson, member of The Beach Boys, died 1998, age 51

AIDS

John Curry, ice skater, died 1994, age 44
Ulysses Dove, choreographer, died 1996, age 49
Elizabeth Glaser, actress and AIDS activist, died 1994, age 47
Frank Israel, architect, died 1996, age 50
Freddie Mercury, singer, died 1991, age 45
Howard Rollins, actor, died 1996, age 46 (AIDS suspected)
Ray Sharkey, actor, died 1993, age 40
Randy Shilts, author, died 1994, age 42

Remember, these are only those who died. The list of young adult victims who are still alive is considerably longer. For example, CFS is fatal only in rare instances, but it has also claimed a number of famous sufferers, including some from the entertainment world like Cher, Randy Newman, and Meshach Taylor.

While the Baby Boom generation is the centerpiece of the book, it is important to note that radiation affects human genetic material, which may harm not only the one directly exposed, but that person's descendants. Chapter 9 discusses the generation born after 1983, who are virtually all children and grandchildren of Baby Boomers. This youngest group of Americans is suffering from growing levels of immune-related conditions similar to those experienced by the Boomers, plus some new diseases that public health officials have just recently begun to measure. Again, multiple factors may all contribute to the decline in health, but because of the poor record posted by Baby Boomers, the effects of radiation on genetics, the continued operation of nuclear power plants, and accidents such as Chernobyl and Three Mile Island, exposures to nuclear products should be considered.

The youngest group of Americans is making negative progress in a number of important, immune-related areas of health status. Childhood cancer incidence is on the rise, as are low-weight births, premature births, asthma, bronchitis, fetal alcohol syndrome, inner ear infections, hypothyroidism, septicemia, measles, meningitis, shigellosis, and whooping cough.

Again, behind each statistic are human stories of suffering, particularly heart-wrenching because they affect small, innocent beings who have barely had the chance to taste life. Anyone who has walked through a pediatric cancer ward of a hospital, as I have, has felt the chill caused by casting eyes on the faces of the afflicted. "Cancer is an obscenity that should happen to no one, much less a child," writes Diane Komp. A pediatric oncologist at the Yale University School of Medicine, Komp describes a number of childhood cancer victims, including one with a form of lung cancer:

> "Little Joe" had been diagnosed with a rare form of
> cancer. Despite all our efforts, the cancer kept coming

> back in his lungs. Now his chest was filling up with
> fluid, and he could not breathe without help. There
> was room for no more radiation, and we knew of no
> other effective treatment. Chest tubes could only drain
> the fluid for a little while before it would reaccumulate.
>
> Our pediatric surgeons were willing to try to strip off
> the lining around the lung that seemed to be "weep-
> ing" in response to the cancer, but the procedure was
> risky. He might not come though the operation at all.
> Even if he did, he might not be able to breathe on his
> own and would forever need the use of a respirator.[12]

Little Joe survived the surgery but, as Komp predicted, continued to
require the use of a respirator. In addition, a breathing tube had to be inserted
into his trachea, robbing the boy of the ability to speak.

Perhaps even more upsetting than the image of a child with cancer is a
premature, underweight baby. The agonizing ordeal of these tiny humans
and their loved ones may be hidden by the high-tech interventions used to
treat them. Many assume that these infants, despite their slow start, will
improve in a Neonatal Intensive Care Unit (ICU), and will often survive, be
discharged, and go on to lead a normal life. This scenario is true in more
and more cases. Science is having greater success than ever in saving premature newborns, or babies with complications at or shortly after birth.

But behind the wizardry of neonatology is a sobering set of facts. A low-
weight birth is much more likely than a newborn weighing over 5½ pounds
to have a complication of delivery, such as fetal distress or cord prolapse.
Underweight babies are far more likely to have an abnormal condition such
as anemia or fetal alcohol syndrome; to require assistance, such as ventilation; to be born with an anomaly such as Down's syndrome, club foot,
hydrocephalus, or malformed genitalia; and finally, is much more likely to
die in infancy. Writer Elizabeth Mehren, age 40, painfully described the plight
of her daughter Emily, born three months prematurely and weighing less
than two pounds, in the spring of 1988. After struggling for six weeks in the
neonatal ICU, amidst a blur of X-rays, blood transfusions, nutritional feed-
ings, and breathing tubes, little Emily was wheeled into surgery to explore
the damage caused by necrotizing enterocolitis, a destruction of the bowel
wall. Elizabeth and her husband Fox waited hopefully for hours, only to
receive devastating news from the surgeon:

> "I'm afraid it couldn't be much worse…her entire in-
> testinal tract was wracked with necrosis. Just de-
> stroyed. There was almost no tissue left alive, nothing
> to work with. We tried to piece it together, but there
> was not enough viable tissue to make it work. There's
> almost nothing left in there."

"What we can do," Dr. Friedman said, "is try to make her as comfortable as possible."

"When will this happen?" Fox asked. "I mean, how long do you think she has?"

"That's impossible to say," Dr. Friedman said. "But usually in a situation like this, death comes within twenty-four to forty-eight hours. For your sakes, I hope its merciful and swift."[13]

Little Emily held on for several more days, bloated by edema, in pain despite morphine, and turning a dusky color, before she died.

While the knowledge linking immune disease to harmful radioactive chemicals has grown through the years, there has been a corresponding lack of action by public health authorities to prevent unnecessary exposure to radiation, raising risk of diseases like cancer. An inordinate amount of energy, time, and money is spent in diagnosing and treating those already suffering from disease, but relatively little is done to *prevent* the disease in the first place. Clearly, the public health system has failed to do its job in protecting the health of the public from carcinogens and other immune-impairing threats.

The steps taken by health authorities to protect the public from all manmade industrial threats have been few and far between, and agonizingly slow. Not until the famous U.S. Surgeon General's report of 1964 did the federal government make a strong statement about the hazards of tobacco use and begin to take action, after decades of silence in the face of widespread use of tobacco in the U.S. and publication of numerous studies showing a link to cancer. It wasn't until 1972, nearly three decades after the introduction of the deadly pesticide DDT, that the national government banned its use. Chlorofluorocarbons (CFCs) were curbed in 1990, but not until they had distinctly weakened the ozone layer of the atmosphere. These are the "success stories"; there are many other offending products that the public health system has taken no action against.

Perhaps the reason behind the failure of public health authorities to adequately protect people from unnecessary immune harm is its unwillingness to take on the industries that create these pollutants. These powerful corporations often hold a strong sway over elected officials, who are fearful that banning toxic substances will hurt profits of wealthy contributors, make it difficult and costly for business to operate (increasing inflation and unemployment), and most of all hurt their chances for re-election to office. Naturally, the will of elected officials is passed down to the health officials they appoint. Some authorities may want to regulate these harmful substances, but fail to act because public agencies do not have the resources to match the corporations in the legal battles that invariably follow. Even the tobacco industry, which is probably the group that public officials have leaned on most heavily in the past, has emerged relatively unharmed. Reduced income from the smaller number of smokers in the U.S. has generally been offset by increased sales overseas. Tobacco growers continue to receive ample subsidies from government. In the more than 30 years since the original Surgeon General's report, in a climate of strong, steady public sentiment against smoking, all but a handful of lawsuits against tobacco companies have been dismissed; and in only two of them were companies held liable, with neither costing the defendants a dime in restitution to the victims so far. The 1997 agreement between the states and tobacco companies will eventually prove to be the first large-scale restitution made by cigarette makers.

The public health establishment's obstinate blindness to recognize and reduce man-made industrial products behind the onslaught of immune disease is perhaps best embodied in a 1995 report issued by the National Cancer Institute, the most prominent body in the nation in the fight against cancer and its causes. The NCI researchers analyzed the spiraling trends in cancer over the past several decades and concluded that increases were likely due to four factors: smoking, increased exposure to the sun, the AIDS epidemic, and improvements in cancer screening and early diagnosis.[14] This conclusion shows that, with the exception of tobacco, which has become the only industrial "whipping boy" that public health officials dare challenge, industry is completely exonerated from any complicity in causing cancer rates to rise. Instead, a "blaming the victim" approach is substituted. People choose to smoke or sit in the sun, and some argue AIDS is due to lifestyle choices, thus blaming people for their own cancers. The credit given to better detection methods hides a powerful silence among a medical profession that has refused to take a public stand against industry-caused cancer.

Chapter 10 addresses the dilemma of proving a thesis that contradicts one that is currently held by many — in this case, the burden of demonstrating that low-level radiation is not harmless, but has and will contribute to extensive sickness and death. Two groups must become convinced that a danger exists, and that steps must be taken to counter the threat. First, the scientific community must understand the relationship, both clinically and statistically, and make appropriate changes in their operations and teachings. However, the scientific community has been generally unwilling to entertain the idea that low-level radiation may be harmful. They are educated in the conclusions first made in the early 1950s that disease incidence in Hiroshima and Nagasaki survivors "prove" low-level exposures aren't damaging to humans. In addition, scientists have long worked under the black cloud of the Cold War and have observed the swift and strong professional retribution against their colleagues who dared to propose and consider these new ideas. The other group that needs to be convinced of a new theory is the public, a constituency perhaps even more important than scientists. In an American-style democracy, the public can raise an outcry for much safer practices and policies, and if this cry is loud enough, it will move elected officials to take action. When this happens, scientists will be forced to comply with new rules. Regulators will force scientists to change and observe safety standards, and researchers who do not believe in the new principle will find it difficult to receive government funding for any work performed in a nonobjective manner. The movement to convince these two groups has had only limited success thus far, but is gaining strength, especially in the post-Cold War era.

Chapter 11 speculates on what may happen to the low-level nuclear threat in the future. There is a mixture of sobering and encouraging news in these concluding pages. On the one hand, there are still 109 nuclear power plants operating in the U.S., an all-time high, producing a daily quota of low-level radioactivity for the environments of millions of Americans. Nuclear waste continues to grow to record amounts, some of it in aging, leaking storage containers, and still there is no long-range plan for its permanent disposal. Nuclear weapons are being dismantled, and no tests have occurred since 1992, but the menace of nuclear war and obliteration still threatens the planet.

On the other hand, no nuclear reactors have been ordered since 1978, and no orders are on the horizon. Some plants have ceased operation in recent years, and the early 21st century will see operators of many other aging reactors decide whether to renovate or shut down. Perhaps the most optimistic news is the evidence that, despite the nuclear horrors of the past, society's members *can* create a healthier society, either by encouraging that unsafe nuclear reactors be closed, or by boosting their own immune systems.

The book provides a solid basis for deciding which direction the nuclear component of American society will take in the future. Acknowledging the past and present damage created by exposure to radioactivity provides the basis for sounder nuclear policies of the future so that to the greatest extent possible, the destruction of the past will not be repeated.

References

1. Epstein, S., *The Politics of Cancer*, Anchor Books, Garden City, NY, 1979.
2. Stewart, A., Webb, J., and Hewitt, D. A survey of childhood malignancies, *British Medical Journal*, June 28, 1958, 1495-1508.
3. Pauling, L., *No More War!* Dodd, Mead, and Company, New York, 1958.
4. Sakharov, A., *Memoirs*, Alfred A. Knopf, New York, 1990.
5. Petkau, A., Effects of 22-Na+ on a phospholipid membrane, *Health Physics*, 239, 1972.
6. Smith, G., As time runs out, *Sports Illustrated*, January 11, 1993, 13-14.
7. Wadler, J., *My Breast: One Woman's Cancer Story*, Addison-Wesley, Reading, MA, 1992, 119.
8. Shilts, R., *And the Band Played On*, Viking Penguin, New York, 1988, 602-3.
9. Smith, M., An American Hibakusha, in *1 in 3: Women with Cancer Confront an Epidemic*, Brady, J., Ed., Cleis Press, Pittsburgh, 1992, 111.
10. Berne, K., *Running on Empty: Chronic Fatigue and Immune Dysfunction Syndrome*, Hunter House Inc., Alameda, CA, 1992, 22, 58.
11. Bell, D., *The Disease of a Thousand Names*, Pollard Publications, Lyndonville, NY, 1991, 22-3.

12. Komp, D. M., *A Child Shall Lead Them: Lessons in Hope from Children with Cancer,* Zondervan Publishing House, Grand Rapids, MI, 1993, 79-86.

13. Mehren, E., *Born Too Soon: The Story of Emily, Our Premature Baby,* Doubleday, New York, 1991, 239-42.

14. Devesa S., et al., Recent cancer trends in the United States, *Journal of the National Cancer Institute,* February 1, 1995, 175-182.

chapter two

The '40s and '50s:
hot war and cold war

The discovery of radioactivity is just over a century old. November 8, 1895, was a Friday, one that saw Wilhelm Roentgen working alone late into the evening in his lab at the University of Wurzburg in Germany. Roentgen, who had been conducting experiments with cathode ray tubes, was testing the density of his tube cover by passing a discharge through the tube, and noticed a faint light on a bench nearby, but well beyond the tube. Roentgen's curiosity was immediately aroused: "I have discovered something interesting, but I do not know whether or not my observations are correct," he remarked soon after.[1] In the next few weeks, Roentgen found this new ray could penetrate not just the tube cover, but all substances — with the exception of lead.[1]

After Roentgen submitted a paper on December 28, the new ray became the object of study by a number of his peers. In February 1896, Henri Becquerel, a physics professor in Paris, followed Roentgen's discovery by showing that exposure to light was not necessary for these new rays; instead, it was the presence of uranium that gave the rays their ability to penetrate substances. Becquerel didn't know it, but he had uncovered the concept known as radioactivity, a term in common use by 1899.[2] The rays quickly became known as Roentgen rays, Becquerel rays, or simply X-rays. Marie Curie, a physics doctoral student in Paris, extended the concept of radioactivity. In 1898, Curie identified another element (thorium) as having the ability to create these strong rays and followed this by announcing two new radioactive elements found in pitchblende ore (radium and polonium, named for Curie's native Poland).[3]

As the world now knows, radioactive elements contain highly charged particles that emit intense concentrations of energy. While the early pioneers did not initially understand this concept, there was no doubt that the new rays had harmful properties unlike anything ever seen. Becquerel, after carrying a tube containing radium in his pocket for a short time, noticed that the skin beneath the pocket had become red and irritated. "I love this radium, but I've got a grudge against it," was his reaction. In July 1896, a man named Hawks became the first victim of X-rays while demonstrating a machine in New York City's Bloomingdale's store. After a period of time, Hawks noticed his nails had stopped growing, his skin became dry, his hair started falling out, and he was in frequent pain.[1] Conversely, scientists also saw unique potential benefits from the new discovery: in June 1901, Curie's husband and colleague Pierre began to speculate that radium could be used to reduce and even eliminate certain tumors.[3]

Quickly, X-rays became part of mainstream medicine in the western world. By 1910, X-rays were being used to diagnose disorders of the digestive tract, and by 1929, organs such as the kidney, ureter, and bladder were also being successfully X-rayed. In 1926, a radioactive sodium iodide solution injected into the carotid artery in the neck made it possible for scientists to better detect brain tumors.[4] Despite these advances, radioactive substances also harmed humans; in 1911, the first reports of X-ray-induced cancer and leukemia in physicians were published. Perhaps the best-known victims of radioactivity in the early years of the century were painters of luminous watch and clock faces. After repeatedly licking the tips of their uranium-coated brushes, a number of workers died prematurely in the 1920s. Subsequent autopsies in Montreal, New Jersey, and elsewhere, confirmed high levels of radioactivity in the deceased workers' bodies, many of whom had died of cancer, and precautions were taken to protect workers from future harm.[4] In 1928, the first official safety limit of radioactivity was set by the International Congress of Radiology meeting in Stockholm, Sweden. The initial annual limit set by the Congress totaled 72 roentgens, a measure of radioactivity named after its discoverer.[5]

Paralleling the advances in medical radiation uses in the first four decades of the century was a continuum of discoveries on the nature of radioactivity. Ernest Rutherford, an Englishman working at Montreal's McGill University, was perhaps the first to anticipate the enormous potential power of radioactivity. In 1903, he wrote:

> The energy of radioactive change must therefore be at least twenty thousand times, and maybe a million times, as great as the energy of any molecular change.[6]

Of the many breakthroughs of these early years, one that occurred in 1932 was called "the turning point in nuclear physics," by renowned physicist Hans Bethe. In February of that year, Manchester University professor James Chadwick found that when radioactive beryllium bombarded other elements, the beryllium's penetration was caused by its neutrons.[6] Chadwick's revelation won him the 1935 Nobel Prize in physics. In March 1934, Italian scientist Enrico Fermi found that bombarding uranium with neutrons produced a new element (later called neptunium), and in March 1941, University of California–Berkeley's Glenn Seaborg discovered plutonium in a similar process.

Clearly, science was sprinting toward a new dimension of knowledge surrounding the properties of the atom. This advancement raised exciting new possibilities for the world, until the technology unfortunately ran head-long into some of mankind's more sinister impulses.

In 1924, a middle-aged man sat alone in a dank Munich prison, contemplating the course of his life and of the world. Adolf Hitler had been, in order, a below-average student, a mediocre painter, and an unexceptional World War I soldier. After the war, Hitler's fanatical German nationalism had led him to organize a political party, but this had also gone nowhere; in 1923, the now-famous "putsch" in a Munich beer hall had backfired, scattering Hitler's small band of renegades and landing him in jail. Now, at age 35, this marginal man addressed the crossroad in his life by resolving himself to his beliefs, and planned the course he would pursue after leaving prison. In his autobiography *Mein Kampf*, written during his prison term, Hitler made clear his ambition of dominating the world:

> In the end only the urge for self-preservation can conquer. The stronger must dominate and not blend with the weaker, thus sacrificing his own greatness. It is no accident that the first cultures arose in places where the Aryan, in his encounters with lower peoples, subjugated them and bent them to his will. If the German people had possessed that herd unity which other peoples enjoyed, the German Reich today would doubtless be mistress of the globe.[7]

The well-known story of Hitler's rise began after his release from jail in 1925 and climaxed with his assumption of dictatorial power in Germany in 1933. By 1939, after disempowering, imprisoning, or killing his political foes, abolishing all freedoms, and making a potent military the centerpiece of German society, the Nazi leader was poised to fulfill his vision of German dominance. On August 22, 1939, on the eve of the first campaign of World War II (the invasion of Poland), he gathered his generals at his Bavarian retreat of Berchtesgaden and shrieked out his win-at-all-cost credo:

> The most iron determination on our part. No shrinking back from anything. Everyone must hold the view that we have been determined to fight the western powers right from the start. A life-and-death struggle. I shall give a propagandist reason for starting the war — never mind whether it is plausible or not. The victor will not be asked afterward whether he told the truth or not. In starting and waging a war it is not right that matters, but victory.[7]

Hitler played an integral role in the history of the nuclear era. First, by threatening the world with aggression and possible enslavement, he motivated his enemies to ensure the destruction of Nazi Germany, no matter what the cost in money and human life. As Albert Einstein said after World War II ended: "If I had known that the Germans would not succeed in constructing the atomic bomb, I would never have lifted a finger.[8] Second, the Nazis were one of the entrants in the race to find a way to introduce nuclear energy to warfare. After receiving a report from German scientists in February 1942, propaganda minister Joseph Goebbels wrote in his diary:

> Research in the realm of atomic destruction has now proceeded to a point where its results may possibly be made use of in the conduct of this war. Tremendous destruction, it is claimed, can be wrought with a minimum of effort. It is essential that we (Germany) be ahead of everybody, for whoever introduces a revolutionary novelty into this war has the greater chance of winning it.[9]

Third, Hitler's extremism caused a brain drain that helped tip the balance in the race for the atomic bomb. On May 10, 1933, thousands of students turned out in Berlin to participate in a public burning of 20,000 books. One fourth of German university professors lost their positions, and university enrollment fell by over half between 1933 and 1939; standards were slackened, and curricula changed to emphasize Nazi priorities.[7] Students sporting the trademark Nazi brown shirts began spreading anti-semitic propaganda in the form of leaflets and fiery speeches, causing an exodus of professors, including many Jews who were starting to feel the persecution from the anti-semitic laws enacted by Hitler's followers. The escapees included some of the greatest minds in nuclear physics, many of whom immigrated to the U.S. during the 1930s from Germany and bordering European nations. This honor roll of refugees included Albert Einstein (1933), Leo Szilard (1933), Edward Teller (1935), Hans Bethe (1933), and Enrico Fermi (1938).[8]

As the Nazi threat grew, the U.S. government made a tentative commitment to exploring whether massive atomic energy could be converted into a military weapon. In the summer of 1939, Albert Einstein and Leo Szilard composed a letter to President Franklin Roosevelt, which they entrusted to Szilard's friend and presidential consultant Alexander Sachs. On October 11, 1939, after meeting with Sachs, Roosevelt remarked "Alex, what you are after is to see that the Nazis don't blow us up. This requires action." However, U.S. progress toward a nuclear weapon was modest until France fell to the Germans in June 1940; and an all-out effort began only after the Japanese attack on Pearl Harbor on December 7, 1941.[9]

The U.S. bomb program, known as the Manhattan Project, comprised three sites. Los Alamos National Laboratory in New Mexico served as the research and planning headquarters for the project. Oak Ridge, Tennessee, formerly a stretch of farmland 20 miles west of Knoxville, became the center of bomb-grade uranium development after October 1943. Hanford, Washington, in the southeast corner of that state, served as the center for plutonium production after September 1944. Even before the outset of the program, the health effects of producing nuclear materials were a prime concern. Major General Leslie Groves, who commanded the Manhattan Project for the U.S. Army, selected the more remote Hanford over Oak Ridge for plutonium production. Any accidental release of the deadly chemical near a major population center such as Knoxville could be disastrous, Groves believed.[6]

When operations at Hanford and Oak Ridge began, it was clear that nuclear fission products being added to the environment posed a hazard. At Hanford, for example, the process of dissolving uranium slugs was only performed when the wind blew away from the major population center of the region.[10] William Wright, an engineer at Hanford during the World War II years, recalls:

> If we started dissolving and the wind got bad, we
> would have to quit, so we were at the vagaries of the
> wind. That's before we had sand filters and all kinds
> of purifiers. The radiation danger was always with us.
> In the early days, we carried Geiger counters with us,
> we didn't have fancy pencils and badges.[10]

Lack of filtration in the stacks of Oak Ridge and Hanford reactors caused large amounts of radiation to be released into the environment. Between 1944 and 1947, for example, Hanford released 685,000 curies of radioactive iodine, a byproduct of plutonium production that damages the thyroid gland, into the air.[11] Perhaps no one knew it at the time, or perhaps the goal of producing the bomb overrode any concern for health, but these emissions were enormous. As a comparison, about 14 curies of iodine escaped during the Three Mile Island accident in 1979. Oak Ridge, which operated for five years with no filtration system in the stacks of its processing plants, also released enormous amounts of radioactivity, including 64,200 curies of iodine in 1947.[12]

These releases were considered part of normal operations, but in addition, early breakdowns in production caused unexpected, accidental releases of radioactivity into the environment. In late 1944, for example, Oak Ridge's pipes in the thermal diffusion plant "leaked so badly, they had to be welded," unleashing uranium gas into the atmosphere. At Hanford, technical problems in the "D pile" operation caused the 1945 release of a large amount of iodine-135 and xenon-135.[6]

Any concern over health effects, however, took a back seat to the principal goal of creating an atomic weapon. Before the first atomic detonation (a test bomb fueled by plutonium and exploded in the remote New Mexico desert near Alamogordo on July 16, 1945), chief safety planner James Nolan commented:

> Possible hazards were not too important in those days. There was a war going on... (Army) engineers were interested in having a usable bomb and protecting security. The physicists were anxious to know whether the bomb worked or not and whether their efforts had been successful. Radiation hazards were entirely secondary.[5]

The Alamogordo bomb was successfully exploded at 5:30 a.m. local time, sending a powerful fireball to a height of 40,000 feet and a mile wide in the soon-to-be-famous form of a mushroom cloud. Scientists tracking the movement of the cloud observed that it first traveled south, then east, splitting into smaller collections of fallout going in three different directions from the variable winds, the largest of which moved to the northeast, cutting a swath 30 miles wide. By the afternoon of July 16, radioactive particles were falling from the sky onto towns as far as 120 miles from the test site.[13] The fallout didn't just stop within New Mexico's borders. In the fall, Kodak manufacturers began to notice that some of the film it had produced came out ruined and unusable. Tracking its steps in manufacturing, Kodak realized that radioactive cerium-141, one of the dozens of radioactive products generated at Alamogordo, had made its way into the Iowa and Wabash Rivers in Indiana and Iowa, where paper for Kodak film was produced.[14]

The story of the two atomic bombs detonated at Hiroshima and Nagasaki has been chronicled many times over. The immediate health effects of the weapons occurred from the intense energy released from the bombs, in effect "vaporizing" many Japanese, and from the acute radiation sickness suffered by those who survived the attacks. By the end of 1945, about 210,000 in the two cities had died from the bomb's effects, and the number rose to 350,000 by the end of 1950.[6] However, this initial focus on the immediate consequences ignored other types of health effects, including Japanese survivors consuming a diet contaminated by radiation, and Americans also eating and drinking food tainted from weapons production.

After the war ended, U.S. nuclear policy could have taken any one of a number of paths. Any question, however, of an American commitment to an all-out nuclear program was settled by the actions of Joseph Stalin's regime in Moscow. Even before the war ended, Stalin recognized the chaos that much of the world was in and viewed the changing balance of power as an ideal opportunity to extend his iron hand. This desire was not a one-time power grab, but a reflection of the essence of Stalin himself. Long before, in 1923, a dying Vladimir Lenin wrote an ominous warning in his journal:

> Stalin has unlimited authority concentrated in his
> hands, and I am not sure whether he will always be
> capable of using that authority with sufficient cau-
> tion.[15]

By 1927, Stalin had forged undisputed control over the Soviet Union,
banishing opponents such as Leon Trotsky, Grigori Zinoviev, and Lev Kame-
nev from the party or the country. Repression of the Soviet people under
Stalin was swift and almost unthinkably cruel. First came the massive famine
of the 1930s, the result of the ill-advised collectivization of the farms, which
was observed coldly by a Stalin who refused to change policy. Between 1931
and 1941, about 9 million Soviet men, women, and children starved to
death,[15] a fact cagily kept secret from the outside by Stalin's henchmen. Next
came the famous purges of the late 1930s, dooming many political, military,
academic, professional, and artistic people to death or imprisonment. Of the
19 million people arrested in the late 1930s, 7 million were executed or died;
and a portion of the rest perished in prison camps.[15] The constant force
driving these savage atrocities was Stalin's paranoia. Nikita Khrushchev, a
Stalin operative during the 1930s and 1940s, described him as

> ... a very distrustful man, sickly suspicious...every-
> where and in everything he saw "enemies," "two-fac-
> ers," and "spies."[15]

At the end of World War II, in which an additional 20 million Soviet
soldiers and civilians died, Stalin took action. During the Yalta conference
in February 1945 with U.S. President Franklin Roosevelt and British Prime
Minister Winston Churchill, Stalin pressed for a redrawing of the borders of
Poland, shifting it west. Over protests, mostly from Churchill but also from
a war-weary and gravely ill Roosevelt, Stalin remained adamant:

> The Prime Minister has said that for Great Britain the
> question of Poland is a question of honor. For Russia
> it is not only a question of honor but of security...a
> matter of life and death for the Soviet state.[9]

When Stalin agreed to create a democratic government in Poland, wel-
coming Polish leaders living in exile in London, the U.S. and Britain gave in.
However, no free elections were ever held, and the U.S.S.R. installed a puppet
government, subservient to Moscow and dominated by the military. This
action was followed by similar maneuvers in all the eastern European
nations that the Soviets had liberated from Nazi rule. By the time Churchill
made his famous "Iron Curtain" speech in 1946, the Cold War was officially on.

Despite the ominous threat posed by Stalin, including an ongoing suspicion that the Soviets were developing their own nuclear weapons, the

American program did not proceed with breakneck speed during the late 1940s. In 1947, the U.S. had few nuclear weapons on hand, and were adding devices only slowly. By mid-1949, Oak Ridge, Hanford, and Los Alamos had been joined only by the Mound production facility in Ohio, plus several research facilities.[16] Between 1946 and 1950, only five weapons tests were conducted, on the Bikini and Eniwetok atolls of the Marshall Islands in the Pacific Ocean. Each of the five devices was between 18 and 49 kilotons (i.e., the equivalent of 18,000 to 49,000 tons of TNT, or one to three Hiroshima/Nagasaki bombs).[17]

America emerged from the atomic bombings in Japan somewhat unnerved by the incredible power of atomic energy, a feeling cushioned by an optimism that atomic energy could be used for the peaceful purpose of producing electric power along with continued advances in medical diagnosis and treatment. Scientists began to speculate about the possibilities of a nuclear energy program, and the newly-created U.S. Atomic Energy Commission (AEC) enlisted the Westinghouse Corporation in 1948 to build an experimental reactor.[18] However, doubts over the feasibility of an atomic power program that would fulfill the dream of "cheap, clean energy" continued. Just weeks after Hiroshima and Nagasaki, General Groves pronounced the commercial use of atomic energy to be "decades" in the future, a belief seconded by other experts soon after.[19]

Another aspect of American optimism regarding atomic energy in the late 1940s coalesced around the hope that world leaders would cooperate in the control of nuclear energy to strengthen the chance that Hiroshima and Nagasaki would never re-occur. Domestically, Congress passed legislation to transfer control of the nation's atomic energy program from the military to civilians, in the form of the AEC, even though the military's presence in atomic matters remained strong. The law was passed over vociferous protests from the military, but countered by numerous scientists who lobbied intensively for civilian control. Also in 1946, the *Bulletin of the Atomic Scientists* was founded to enlighten the public about atomic energy. In the late 1940s, finally free of the shackles of great depression and world war, the American public tended toward an enjoyment of the new order, and shied away from serious issues like atomic bombs. A concerned Einstein identified this apathy in 1947:

> The public, having been warned of the horrible nature
> of atomic warfare, has done nothing about it and to a
> large extent has dismissed the warning from his con-
> sciousness.[8]

On the international front, hopes were set on the possibility of cooperative multi-national control of nuclear energy. In late 1945, a hot debate began over whether the U.S. should pass information to a body supervising world control of nuclear energy. The proposal drew a number of supporters including government figures Henry Stimson, Henry Wallace, Eleanor Roosevelt, and J. William Fulbright; scientists Harold Urey, Albert Einstein, Vannevar Bush, James B. Conant, Eugene Wigner, and J. Robert Oppenheimer; and the governments of many western nations. However, President Harry Truman and most U.S. congressmen remained firmly against the idea, along with virtually all military leaders, led by an outspoken Groves.[20] The proposal was bandied about in the United Nations for about two years before talks broke off in 1947, with no report being sent to the Security Council.

Almost unnoticed in discussions of nuclear policy in the late 1940s was the matter of health effects of radiation exposure. As mentioned, the focus on study of the Hiroshima/Nagasaki population had been *entirely* on the short-term effects of a *single high dose* of radiation, rather than long-term effects of lower dose exposures. No investigation was made of the effects of ingesting radioactivity through the food chain. The Alamogordo blast, known as Trinity, also produced little interest. The Oak Ridge National Laboratory scrutinized cattle that received a large dose of radiation from Trinity, observing that the radioactive dust "turned the hair on their sides and backs from deep red to gray or white and left spots on their skin which looked very much like burns." However, James Kile, the chief veterinarian on the government's Oak Ridge team, pronounced their condition as "good" and proclaimed the milk produced by the cows to be "not radioactive milk. Just milk."[21]

Amidst the prevailing silence, a few scant voices spoke out about adverse health effects of radiation in those early years. In January 1947, British geneticist J.B.S. Haldane told a conference of scientists that the greater danger in the atomic bomb lay not in its immediate destruction, but in the subsequent genetic damage to survivors that is passed down to many generations.

> The tremendous amount of radiation generated in the explosion of an atomic bomb produces mutations in the genes, carriers of heredity. These mutations in the survivors of Hiroshima and Nagasaki will affect future generations. The killing of 10 percent of humanity by an attack with atomic bombs might not destroy civilization. But the production of abnormalities in 10 percent of the population by gene mutations induced by radioactivity may very easily destroy it.[22]

The following year, UCLA's Albert W. Bellamy predicted that radiation might kill as many as 9 of 10,000 offspring of the generation affected by an atomic bomb, due to genetic disorders. Bellamy added that nonfatal afflictions such as hemophilia and feeble-mindedness would increase, compounding the genetic effects of the bomb.[23] For the most part, however, voices such as Haldane and Bellamy went unnoticed. The AEC, while admitting that U.S. nuclear operations were harmless to people, nonetheless took a number of steps reflecting concerns over health and safety. In 1947, the AEC created a Health and Safety Lab in New York; formed the Atomic Bomb Casualty

Commission to study effects on Hiroshima/Nagasaki survivors; began conducting research on animals at its Argonne National Lab in Illinois; and requested that the Air Force develop a means of measuring radioactive fallout levels.[24]

The limited early debate on health effects of man-made radiation did not stay limited for very long. While the late 1940s was a time of amazement over this grand new invention, the 1950s became a period in which reality set in and science began to move toward a better understanding of radiation's mechanisms in human biology.

As usual, political events drove science's progress. In the few years after World War II ended, the U.S. nuclear program was generally limited, reflecting an uneasy but relatively stable standoff between the capitalist and communist blocs. Just a few months after Churchill gave his gloomy Iron Curtain speech in 1946, former Vice President Henry Wallace made a speech exhorting the U.S. to pursue conciliation with the Soviets. The Marshall Plan, devised by Secretary of State George C. Marshall, became one of the backbones of U.S. foreign policy in April 1948, in the hope that a rebuilt, vital Western Europe would help guarantee the peace with Soviet bloc countries. The formation of the North Atlantic Treaty Organization (NATO) in March 1949 bolstered the U.S.–Western European pact. While tense events occurred from time to time, none seemed threatening. The 1947 communist insurgency in Greece, Czechoslovakia's loss to communism in February 1948, and the Berlin airlift in 1948 were troubling, and Stalin's brutality and ambitions were well known, but a coexistence appeared to be in place.

Any stability in this coexistence was shattered in the 11 months between August 1949 and July 1950. "A year of shocks," is how historian Eric F. Goldman put it.[25] No less than eight equally disturbing developments conspired to throw the world into turmoil. The first shock was delivered on August 5, 1949, when Mao Tse-Tung's Communists overran mainland China, after a long and arduous civil war. The Soviet government immediately recognized and signed a friendship pact with the new regime, while the U.S. recognized the deposed Nationalists, now exiled to the island of Taiwan. The second, perhaps most unsettling, news flash was delivered on September 23, when Truman announced that the Soviets had successfully exploded an atomic device on August 29, a statement quickly confirmed by Moscow. Truman moved fast and, despite opposition by an expert panel of scientists led by J. Robert Oppenheimer, announced on January 31, 1950, that the U.S. would proceed with development of a hydrogen-based nuclear device, a much more powerful weapon than the uranium- or plutonium-based atomic bomb.

Fear of communism seized the country, reinforced by a continuing barrage of jarring events. On January 21, former State Department official Alger Hiss, suspected of communist ties and espionage, was convicted of perjury in New York City, and began a four-year jail term. Two weeks later, Klaus Fuchs, a scientist who had been on the Los Alamos team in 1945, was captured in London and later confessed to passing secret information to Stalin's government on atomic bomb construction and testing. Just days after, an obscure U.S. Senator from Wisconsin made a speech in Wheeling, West Virginia, which had received virtually no prior buildup; but immediately after he made the charge that communists were infiltrating the State Department, all Americans knew the name of Joseph McCarthy, and a long, agonizing, and eventually fruitless witch hunt for domestic "reds" was under way. The exact text of McCarthy's speech no longer exists, but his notes suggest that the Senator unloaded the following bombshell:

> While I cannot take the time to name all of the men in the State Department who have been named as members of the Communist Party and members of a spy ring, I have here in my hand a list of 205...a list of names that were known to the Secretary of State and who nevertheless are still working and shaping the policy of the State Department.[26]

To cap the "year of shocks," North Korean Communists stormed across the 38th parallel into South Korea on June 24 with approval from Moscow and Peking, marking the start of the Korean War. Finally, Julius Rosenberg was arrested in New York on July 18, accused of passing classified information to the Soviets. His wife Ethel was also arrested the following month, and the two would be convicted and executed in 1953.

The chill of the Cold War permeated every aspect of American society. The relative calm of postwar America had been shattered, and was replaced by an unnerved, unsettled feeling. No public or private institution was spared the suspicion that within its ranks, "reds" bent on taking over the country were lurking. The atomic program of the U.S. government, already enshrouded by a cult of secrecy left over from the war, was not scrutinized nearly as much as other federal targets of McCarthy and his followers, such as the State Department and the U.S. Army. Still, there was some level of communist-hunting directed at the nuclear weapons program, even before activities of spies like Fuchs and the Rosenbergs became known. As early as 1947, Congressman J. Parnell Thomas, chairman of the House Un-American Activities Committee, leveled the charge that "fellow travelers, if not actual members of the Communist party," were active within the Oak Ridge weapons manufacturing complex. This charge provoked an angry response from John H. Bull, executive committee chairman of the Association of Oak Ridge Engineers and Scientists:

> I know of no Oak Ridge Communists, never heard of one before, and have never seen one here.
>
> We challenge Thomas to bring his charges against any leading scientist in Oak Ridge or against any officer or member of the executive committee, present or past, of the Association of Oak Ridge Engineers and Scientists.[27]

Perhaps the defining event of the impact of communist-hunting on the nuclear industry was the saga of J. Robert Oppenheimer. In the late 1930s, Oppenheimer, already a distinguished professor at the University of California at Berkeley, became engaged to Jean Tatlock, an intermittent member of the American Communist party. For the next few years, the previously apolitical Oppenheimer went to an occasional party meeting and contributed $500 to $1000 to the cause each year. By 1941, Tatlock and he were no longer romantically involved, and his dalliance with party activities had ended. During and after World War II, American intelligence kept Oppenheimer under nearly constant surveillance, but because of his brilliant performance at Los Alamos and after, no official questioned his allegiance to the U.S.

The 1950s were a different story. In November 1953, FBI Director J. Edgar Hoover received a letter from William Borden, former head of the Congressional Joint Committee on Atomic Energy. Borden was disturbed by the recommendation of the AEC's General Advisory Council chaired by Oppenheimer against pursuing a hydrogen bomb, advice that was later rejected by Truman. Borden wrote Hoover that Oppenheimer "more probably than not" was an agent of the Soviet Union, acting under Soviet directive. Hoover forwarded the letter to President Dwight D. Eisenhower, who promptly suspended Oppenheimer's security clearance, denying him access to otherwise restricted data in the U.S. nuclear program.[28]

On April 12, 1954, the AEC convened a special hearing of its Personnel Security Board to consider Oppenheimer's case. During the three weeks of proceedings, Oppenheimer declared his strong opposition to communism, and most of the 40 witnesses attested to Oppenheimer's loyalty to the U.S. John Lansdale, the chief security officer for the Manhattan Project, was typical of the majority of witnesses.

> I have never, strongly as I have felt and acted with reference to Communism, adopted the assumption once a Communist sympathizer, always a Communist sympathizer. I do feel strongly that Dr. Oppenheimer, at least to the extent of my knowledge, is loyal. I am extremely disturbed by the current hysteria of the times of which this seems to be a manifestation.[28]

However, a handful of witnesses testified against Oppenheimer, including the zealously pro-nuclear physicist Edward Teller:

> In a great number of cases I have seen Dr. Oppenheimer act...in a way which for me was exceedingly hard to understand. I thoroughly disagreed with him in numerous issues and his actions frankly appeared to me confused and complicated. To this extent, I feel that I would like to see the vital interests of this country in hands which I understand better, and therefore trust more.[28]

In the end, Cold War anxiety won out. The Board found Oppenheimer "exercised poor judgment in continuing some of his past associations into the present," and found "his conduct in the hydrogen bomb project sufficiently disturbing." The Board recommended a permanent denial of access to restricted data, which was upheld by the AEC's 4 to 1 vote.[28]

The American nuclear weapons system was strongly affected by the steady stream of bad news in 1949 and 1950. As mentioned, the introduction of the H-bomb was one major change in policy. The other occurred on December 18, 1950, when Truman decided to begin testing atomic weapons in the continental U.S., in addition to the Pacific. Just 21 months earlier, AEC Commissioner Sumner Pike had declared that "only a national emergency could justify testing in the United States." Now, the AEC carried out Truman's order to develop the Las Vegas–Tonopah Bombing and Gunnery Range, a large plot of desolate land owned by the U.S. government 70 miles northwest of Las Vegas, as a site for bomb testing, in addition to the Marshall Islands. Truman and his advisers chose the site over others in Nevada, Utah, New Mexico, and North Carolina because the low population density and the prevailing westerly winds (blowing away from the heavily-populated west coast) seemed to reduce any potential hazards from bomb fallout.[24]

The first test in Nevada (code named "Able") occurred on January 27, 1951, at 5:44 a.m. local time. Dropped from a B-50 bomber and detonated at an altitude of 1,060 feet, Able had the explosive power of only 1,000 tons, far below that of the Hiroshima (15,000 tons) or Nagasaki (21,000 tons) bombs. Able was followed in the next 10 days by four more bombs dropped from B-50s, with a total of 39,000 tons of explosive power.[17]

The effects of fallout from these tests were felt almost immediately around the country. By the afternoon of January 27, the radioactive cloud formed by Able had reached Missouri, and by the morning of January 29, entered the atmosphere above southern New England as a cloud 300 miles in width. That same morning, as snow fell in Rochester, NY, Eastman Kodak technicians noticed their Geiger counters "clicking wildly" and promptly notified the AEC. Other fallout that day was detected in places like Ithaca, NY, at 10 times the normal level of background radiation, and Lexington, MA, at 5 times the usual amount.[14]

Although newspapers and other media reported on the increased fallout from Nevada, there were rarely any mentions of actual levels of radioactivity, nor were any specific radioactive chemicals making up the fallout "cocktail" discussed. Instead, reporters dutifully relayed the AEC party line that fallout would linger only for a matter of days (ignoring the presence of long-lived elements in fallout) and that increased levels were still far below the government's threshold for harm to human health. This belief went unquestioned for several years, through the Nevada test series of 1951, 1952, and 1953, along with more testing in the Marshall Islands in 1951 and 1952 (including the first thermonuclear, or hydrogen, bomb on October 31, 1952, with the power of 10,400,000 tons of TNT).

In 1953, however, the unquestioned faith of Americans in the government's assurance that bomb test fallout was harmless began to show signs of cracking. Two days after the "Simon" test shot on April 25, a violent thunderstorm hit the Albany–Troy area of upstate New York, bringing fallout from Simon to the ground. A radiochemistry class from Troy's Rensselaer Polytechnic Institute measured radioactivity levels in the tap water, finding them to be well in excess of federal limits. However, as radioactivity dropped over the next few days (to levels still well above normal), the relieved professor of the class turned in his report to the AEC, which kept the information secret.[29] A public uproar had been temporarily avoided, but only after a close call.

That same spring, another alarm arose in Utah. In Cedar City, near and directly downwind from the Nevada test site, sheep began dropping dead while grazing on the hillsides and pastures, soon after the "Nancy" shot on March 24. In all, 4390 sheep died within weeks, prompting complaints from sheep herders and an immediate AEC investigation.[30] Major Robert Veenstra, a veterinarian who examined the dead sheep, along with surviving sheep and lambs, learned that the AEC was deliberately overlooking many of his findings, so he filed a letter with the Commission later in the year:

> The impression seems to exist that I based my opinion regarding the Utah and Nevada livestock problem entirely on the finding of I-131 in the thyroids. This is not the case. As you know, we found radiation present in the kidney, spleen, rib, and liver as well as in the thyroid. In addition, the presence of lesions and their type and location on the animals must not be overlooked.[30]

The Commission was taking no chances, however. At a public hearing in Cedar City on August 9, AEC official Paul Pearson, who already knew of the massive doses of iodine absorbed by the thyroids of the dead sheep, told the farmers that drought might be a factor in the sheeps' demise, that "no one has seen any radiation burns on sheep," and "malnutrition certainly is a major contributing factor."[30] The final AEC report, issued January 16, 1954, "closed the book" on the investigation according to *The New York Times*, even though the sheep herder's case remained open in the courts well into the 1980s. Despite the obvious damage implicating fallout as the cause, the bomb testing went on.

With public concern over the health effects of radiation still bottled up in the early 1950s, the military and its supporters had carte blanche to run the weapons program however they wanted. Many believed that atomic weapons could and should be used, and their talk was often tough. Early in 1951, with the Korean War in full swing, Pennsylvania Congressman James Van Zandt made the following comments in a television interview:

> I've always been a firm believer that we should use the atomic bomb not only on Korea but north of the Yalu River in Manchuria. I think that there are several targets, uh, in northern Korea that we could use, uh, we could use — that is, we could destroy them with the atomic bomb...we could destroy them and contaminate them. And then, of course, there are targets in Manchuria that *should* be destroyed.[14]

The AEC made no effort to conceal why weapon after weapon was being tested. On April 21, 1952, AEC Chairman Gordon Dean gave a press conference before the test shot known as "Charlie," which was so powerful that its fallout particles circled the globe before depositing radioactivity on the west coast:

> Today, atomic weapons are thought of as tactical, as well as strategic weapons — that is — they are thought of as weapons that can be employed by military forces in the field. In other words, they are thought of as weapons which tactical air force and armies and navies — as well as strategic air forces — have legitimate interest in and a legitimate need for.[14]

One opposing viewpoint in this otherwise gung-ho era was the speech given by President Eisenhower before the United Nations on December 8, 1953. Known as the "Atoms for Peace" speech, Eisenhower expressed fear over the growing arms race and extended an olive branch to the Soviets, namely the dual goals of arms reduction and promotion of peaceful applications of atomic energy:

> The U.S. "would seek more than the mere reduction or elimination of atomic materials for military purposes...it is not enough to take this weapon out of the hands of the soldiers. It must be put in the hands of those who know how...to adapt it to the arts of peace...this greatest of destructive forces can be developed into a great boon, for the benefit of all mankind."[31]

The Soviet Union balked at the idea of placing nuclear materials into a pool, fearing it would fall farther behind in the arms race. However, there was a response on the domestic front. In 1954, Congress passed the Atomic Energy Act, which allowed private companies to build nuclear power plants and ordered the AEC to work cooperatively with these firms.[32] On September 6, 1954, ground was broken for construction of the first U.S. commercial nuclear power station, a site near Pittsburgh known as Shippingport.

This time of re-thinking of the consequences of the arms race coincided with the first movements by government and private leaders to call attention to the hazards posed by exposure to bomb fallout. Perhaps the groundwork for this thaw was laid by the Atomic Bomb Casualty Commission, which identified the first delayed reactions to the Hiroshima and Nagasaki bombs. In 1950, a cluster of persons with cataracts in the two cities was identified, followed the next year by a high incidence of leukemia, and in 1954 by the discovery of high rates of miscarriages, stillbirths, infant deaths, and retardation.[33] Despite these findings, it wasn't until 1955, the fourth year of Nevada testing, when health officials spoke out. On March 12, Ray R. Lanier of the University of Colorado's radiology department made a statement to the Associated Press after a large bomb was tested in Nevada:

> For the first time in the history of the Nevada tests, the upsurge in radioactivity measured here within a matter of hours after the tests has become appreciable.[14]

Just days later, the New Jersey Labor and Industry Commissioner said the state faced an "insidious hazard" from Nevada fallout.[34] Despite denials of any hazards by the AEC and U.S. Public Health Service, the first volleys in the crusade to halt atmospheric testing in Nevada had been fired.

The year 1956 can be seen as a key year in the gradual realization by both American and Soviet officials that the arms buildup and the fallout produced by weapons manufacturing and testing was harming people. The first sign of a Soviet thaw came in February 1956, during the 20th Congress of the Communist party in Moscow. Nikita Khrushchev, a former Stalin operative still striving to consolidate his power after Stalin's death three years earlier, strode to the podium and made an impassioned speech:

> When we analyze the practice of Stalin in regard to the direction of the Party and of the country, when we pause to consider everything which Stalin perpetrated, we must be convinced that Lenin's fears were justified. The negative characteristics of Stalin, which, in Lenin's time, were only incipient, transformed themselves during the last years into a grave abuse of power by Stalin, which caused untold harm to our Party. It is clear here that Stalin showed in a whole series of cases his intolerance, his brutality, and his abuse of power. Instead of proving his political correctness and mobilizing the masses, he often chose the path of repression and physical annihilation, not only against actual enemies, but also against individuals who had not committed any crimes against the Party and the Soviet government.[35]

On the other side of the world, Eisenhower was facing a challenge from Illinois Senator Adlai E. Stevenson in his bid for re-election as President. For most of the campaign, nuclear weapons got little attention aside from the standard Cold War mottoes of national strength and containment of Communism. No mention was made of bomb testing and fallout hazards. However, on October 15, Stevenson made a bold move. Trailing badly in the polls, he made a speech in Chicago televised nationally by ABC, in which he called for a pact to suspend hydrogen bomb testing, specifically citing the hazards created by fallout:

> With every explosion of a super-bomb huge quantities of radioactive materials are thrown up into the air — pumped into the air currents of the world at all altitudes — and later on they fall to earth as dust or in rain. This radioactive fall-out as it is called, carries something that's called strontium-90, which is the most dreadful poison in the world. For only one tablespoon equally shared by all the members of the human race could produce a dangerous level of radioactivity in the bones of every individual. In sufficient concentrations it can cause bone cancer and dangerously affect the reproductive processes.[36]

The 1956 moves by Khrushchev and Stevenson had little immediate impact. For all their novelty, the bold oratory by Khrushchev did not alter Soviet nuclear policy. Although his denunciation of Stalin officially ended the worst of the Soviet repression (one reason why the nuclear arms race had heated up), Khrushchev assumed dictatorial powers and carved out his own niche as a passionate Cold Warrior. Stevenson's statement was repudiated by Eisenhower ("the continuance of the present rate of H-bomb testing...does not imperil the health of humanity"),[37] who easily won re-election and resumed America's testing program. Nonetheless, the seeds for the end of bomb testing were being sown in both Washington and Moscow.

Meanwhile, reports of rising fallout levels were becoming more frequent and more alarming. On November 1, 1956, with Stevenson's campaign in its final days, Senator Hubert H. Humphrey told a group of midwestern farmers that strontium-90 in Wisconsin and Illinois soil had quadrupled from 1953 to 1955.[38] That same day, New York Governor Averill Harriman reported that radioactivity in his state in late September had risen to levels 10 to 60 times above normal.[39] In the spring of 1958, a page one story in *The New York Times* showed the full extent of fallout penetration into the environment. An extensive article was accompanied by a chart showing the level of strontium-90 in New York City to be 2 millicuries per square mile at the beginning of 1954; the figure rose to 7 in 1955, 16 in 1956, 28 in 1957, and 43 at the outset of 1958, with the level still rising.[40] The continued denials by the AEC and other public officials that emissions from bomb tests were causing no harm to the public was falling on increasingly deaf ears.

As the 1950s went on, more accepted the belief that low levels of radiation presented its own set of health threats, perhaps distinct from the high doses of a bomb used in warfare. Several brave scientists stepped forward during this time, into the dark shadows of Cold War politics, to express their views. In 1956, a National Academy of Sciences committee issued a report asserting no safe threshold for radiation exposure; that is, even small doses can be harmful. The panel condemned excessive use of medical and dental X-rays, along with X-rays of pregnant women and people being fitted for shoes.[41] Later that year, British physician Alice Stewart published a preliminary communication in the British journal *Lancet* identifying a doubling of leukemia and cancer risk in childhood after the mother had undergone an abdominal X-ray during pregnancy.[42] Two years later, Stewart corroborated her findings in an expanded article that evoked a firestorm of controversy[43] until Brian MacMahon of the Harvard University School of Public Health produced a similar study in 1962.[44] Thus, Stewart had pinpointed for the first time a specific danger of low levels of radiation exposure.

The spirit of this discovery made its way into the area of fallout as well, making the issue of continued bomb testing a public one. Again, towering figures, one each from the U.S. and the U.S.S.R., were instrumental in moving this debate along. Andrei Sakharov was a physicist who had been instrumental in the Soviets' development of the hydrogen bomb. However, after the second detonation of the H-bomb on November 22, 1955, Sakharov surprised his fellow scientists by making a toast: "May all our devices explode as successfully as today's, but always over test sites and never over cities."[45] Sakharov, troubled by the growing health hazards presented by the arms race, began to explore just how much damage was being done. In a 1958 article in the Soviet journal *Atomic Energy*, authorized by Khrushchev even though it went unnoticed in the west, Sakharov calculated there would be 10,000 human victims for each megaton of nuclear explosion. (Because total output of U.S./U.S.S.R. tests from 1945–1963 would eventually reach 600 megatons, Sakharov's formula suggests about 6,000,000 victims worldwide.) In the article, Sakharov wrote:

> Halting the tests will directly save the lives of hundreds of thousands of people, and it also promises even greater indirect benefits, reducing international tension and the risk of nuclear war, the fundamental danger of our time.[45]

Sakharov's American counterpart was Nobel Prize winner Linus Pauling. Like Sakharov, Pauling projected enormous suffering and loss of life because of bomb testing. In 1957, he asserted that 1,000,000 persons would lose 5 to 10 years of life if testing continued.[46] He also cited estimations of hundreds of thousands of genetic defects in current and future generations due to bomb tests.[47] Unlike Sakharov, who lived within the restrictions of a totalitarian society that repressed free speech, Pauling was able to bring his message into the public arena. On January 13, 1958, he presented a petition to the United Nations, calling for an immediate halt to all bomb testing. Signed by 9,235 scientists, including 37 Nobel Prize winners, the petition singled out the threat posed by strontium-90 in milk as a major reason behind the proposed testing halt.[47]

Gradually, the efforts of pioneers like Sakharov and Pauling along with the germinating uneasiness among the public had an effect on those in power. Eisenhower, a staunch defender of the belief that fallout presented no health threat, uneasily acknowledged the growing opposition at a June 5, 1957, news conference, grumbling that the resistance "looks like almost an organized affair," and (unconvincingly) adding "we have reduced the fallout from bombs by nine-tenths."[14] With public skepticism over atomic energy growing, the business community became increasingly skittish about developing nuclear power, an aspect of the field that gave military applications more respectability. In 1957, a nervous Congress passed an amendment to the 1954 Atomic Energy Act, limiting corporate liability to $60 million in case of a nuclear accident and total liability to $560 million. The legislation breathed new life into the shaky nuclear power program (at that point, only seven nuclear power plants had been ordered, and none had begun operation). Westinghouse's Charles Weaver summarized what the law meant to an industry uncertain of the safety risks and their consequences:

> We knew at the time that all questions weren't answered. That's why we fully supported the Price–Anderson liability legislation. When I testified before Congress, I made it perfectly clear that we could not proceed as a private company without that kind of government testing.[32]

In 1958, the two superpowers finally succumbed to the mounting evidence of danger and public outcry, and negotiated an agreement to halt all nuclear testing indefinitely. That fall, after a final volley of 13 tests in seven days, the U.S. tested its last nuclear weapon for nearly three years on October 30, 1958, with the explosion of a small device above ground and a larger underground bomb (the U.S. had begun to hold some of its bomb tests in underground tunnels and shafts late in 1957).[17] With the moratorium in place, the concern abated somewhat, although no one was willing to relax in the continuing grip of the Cold War.

References

1. Glasser, O., *Wilhelm Conrad Roentgen and the Early History of the Roentgen Rays*, John Bale, Sons and Danielson Ltd., London, 1933.
2. Romer, A. (ed)., *The Discovery of Radioactivity and Transmutation*, Dover Publications, New York, 1964, 11.
3. Curie, E., *Madame Curie*, DeCapo Press, New York, 1937.
4. Donizetti, P., *Shadow and Substance: The Story of Medical Radiography*, Pergamon Press, Oxford, 1967.
5. Hacker, B., *The Dragon's Tail: Radiation Safety in the Manhattan Project*, University of California Press, Berkeley, 1987, 18, 84.
6. Rhodes, R., *The Making of the Atomic Bomb*, Simon and Schuster, New York, 1986, 43.
7. Shirer, W., *The Rise and Fall of the Third Reich*, Simon and Schuster, New York, 1959, 86-89, 532.
8. Jungk, R., *Brighter Than a Thousand Suns: A Personal History of the Atomic Scientists*, Harcourt, Brace, and Company, New York, 1958, 248.
9. Goodwin, D., *No Ordinary Time*, Simon and Schuster, New York, 1994, 346, 581.
10. Sanger, S. L., *Hanford and the Bomb: An Oral History of World War II*, Living History Press, 1989, 119.
11. Hanford Health Information Network, *The Release of Radioactive Materials from Hanford*, Seattle, 1993, Vol. 2.
12. Oak Ridge Health Assessment Steering Panel, *Oak Ridge Health Studies Phase I Overview*, Nashville, TN, 1993, Vol. II, Part B, E4-E9.
13. Lamont, L., *Day of Trinity*, Atheneum, New York, 1985, 251-2.
14. Miller, R., *Under the Cloud: The Decades of Nuclear Testing*, The Free Press, New York, 1986, 101, 146, 197-198.
15. Lewis, J. and Whitehead, P., *Stalin: A Time for Judgment*, Pantheon Books, New York, 1990, 54, 99.
16. *The New York Times*, May 29, 1949, IV5.
17. Norris, R. and Cochran, T., *United States Nuclear Tests, July 1945 to 31 December 1992*, Nuclear Resources Defense Council, Washington, DC, 1994, 22.
18. *The New York Times*, December 29, 1948, 1.
19. *The New York Times*, August 31, 1945, 4; May 6, 1948, 37.
20. *The New York Times*, various dates in 1945.
21. *The New York Times*, September 22, 1946, VI 58.
22. *The New York Times*, January 4, 1947, 17.
23. *The New York Times*, August 14, 1948, 15.
24. Ball, H., *Justice Downwind: America's Atomic Testing Program in the 1950s*, Oxford University Press, New York, 1986.
25. Goldman, E., *The Crucial Decade — And After, 1945-60*, Vintage Books, New York, 1960, 112.
26. Oshinsky, D., *A Conspiracy So Immense: The World of Joe McCarthy*, The Free Press, New York, 1983, 109.
27. *The New York Times*, June 6, 1947, 3.
28. Wharton, M., *A Nation's Security: The Case of J. Robert Oppenheimer*, Secker and Warburg, London, 1955, 119, 126, 275.
29. Sternglass, E., *Secret Fallout: Low-Level Radiation from Hiroshima to Three Mile Island*, McGraw-Hill, New York, 1981, 1-4.

30. Fuller, J., *The Day We Bombed Utah: America's Most Lethal Secret*, New American Library, New York, 1984, 89.

31. Ambrose, S., *Eisenhower: Soldier and President*, Simon and Schuster, New York, 1990, 342.

32. Hertsgaard, M., *Nuclear Inc.: The Men and Money Behind Nuclear Energy*, Pantheon Books, New York, 1983, 33.

33. *The New York Times*, June 18, 1950, 36; January 31, 1952, 4; May 1, 1954, 36.

34. *The New York Times*, March 15, 1955, 17.

35. Crankshaw, E., *Khrushchev Remembers*, Little, Brown and Company, Boston, 1970, 564, 569.

36. *The New York Times*, October 16, 1956, 18.

37. *The New York Times*, October 24, 1956, 1.

38. *The New York Times*, November 2, 1956, 27.

39. *The New York Times*, November 2, 1956, 54.

40. *The New York Times*, April 4, 1958, 8.

41. *The New York Times*, June 13, 1956, 1.

42. Stewart, A., et al., Malignant disease in childhood and diagnostic irradiation in utero, *Lancet*, September 1, 1956, 447.

43. Stewart, A., Webb, J., and Hewitt, D., A survey of childhood malignancies, *British Medical Journal*, June 28, 1958, 1495-1508.

44. MacMahon, B., Prenatal xray exposure and childhood cancer, *Journal of the National Cancer Institute*, May 1962, 1173-91.

45. Sakharov, A., *Memoirs*, Alfred A. Knopf, New York, 1990, 191, 202.

46. *The New York Times*, June 3, 1957, 11.

47. Pauling, L., *No More War!* Dodd, Mead, and Company, New York, 1958, 12.

chapter three

The '60s and '70s: nuclear progress amidst fear and doubt

In early 1960, nuclear arms and power were not particularly high on the list of public concerns for Americans. Atomic bomb testing by both the U.S. and U.S.S.R. had halted, and negotiations for a treaty banning all tests or just atmospheric tests continued. Although long-lived radioactive products such as strontium and cesium remained in the milk, water, soil, and air, levels began to drop from their late-1950s heights. While optimistic predictions of a great emergence of nuclear-generated electricity abounded, only one plant was in operation, and only 13 others had been ordered by utilities; the American public still saw nuclear power plants as sort of an abstract concept, tied more to the future than the present.

This relative calm didn't have long to live, however. As the decade proceeded, nuclear issues — both military and civilian — assumed great prominence in American life, with human health effects of exposure to radioactive products taking the spotlight. This renewed prominence was made possible by nuclear-related events dominating the times. Moreover, the spirit of the era moved many to vocalize their opinions on nuclear matters for the first time, affecting official policies. The 1960s was a decade of honest reckoning with issues that were often unpleasant to face; a decade in which awareness of environmental protection moved people as never before; and a decade in which the Cold War hit its abyss before reversing its field and unraveling, ever so slowly.

Some believe the 1960s actually began on February 1, 1960. On that unseasonably warm day, four young college students stepped into the F.W. Woolworth's on Elm Street in Greensboro, North Carolina, seated themselves at the lunch counter, and ordered doughnuts and coffee. Normally, this would be a routine, unremarkable act, except that these four men were black, and were seated at the food-service area reserved only for whites (blacks could only use a nearby stand-up snack bar). When they were refused service, the students calmly sat at the counter as word spread, and curious (and sometimes angry) shoppers paused to watch. They returned the following day and the day after that, each time bringing with them a larger contingent of blacks and attracting more attention from the press. A sit-in at a second store in Greensboro began, followed by similar demonstrations throughout the state and the southeast, which had long been dominated by rigidly separatist Jim Crow laws. America had begun its hard look at *de jure* segregation and the broader context of equal rights under the law, a look that was to cause the country considerable agony as the price for progress.[1] Furthermore, the civil rights movement was a populist, democratic movement guided by the masses rather than the elite, setting a precedent for other movements soon to follow.

Two years after the Greensboro sit-ins, with the civil rights movement in full swing, another pioneering effort to help understand the human condition was kicked off. Rachel Carson, a marine biologist best known for books on sea life such as *The Sea Around Us*, published *Silent Spring*, a strong treatise warning of the dangers of pesticides to human health. The book became widely read, and provoked an agitated backlash from the chemical industry. President John F. Kennedy read the book and was reportedly moved by it. Carson's death from cancer in 1964 only extended the interest in the book, in the effects of pesticides, and in the overall plight of the environment.

The 1960s' spirit of public involvement in important issues, the growing fear that mankind was poisoning itself, and leftover uneasiness from the initial years of the nuclear era in the 1940s and 1950s now converged to set the stage for a decade of introspection and change for atomic energy. All that was needed was an event to kick-start this process; the event occurred right at the beginning of the decade.

On May 1, 1960, U.S. Air Force pilot Francis Gary Powers floated across the sky, about 68,000 feet over the Soviet city of Sverdlovsk. Powers was piloting a U-2 spy plane flight from Peshawar, Pakistan, to Bodo, Norway, a 3800-mile journey lasting about nine hours. U-2 flights had been taking place since 1956, but this was the first attempt to fly straight across the Soviet Union and continue on to another landing site, rather than fly over part of the U.S.S.R. and then return to the original air base.

At 10:30 a.m. local time, four hours into the flight, flying high enough to escape detection from ground surveillance, Powers suddenly ran into trouble. His recollection did not do anything to clarify whether the plane had been fired on or had simply malfunctioned:

> Following the turn, I had to record the time, altitude, speed, exhaust, gas, temperature, and engine–instrument readings. I was marking these down when, suddenly, there was a dull "thump," the aircraft jerked forward, and a tremendous orange flash lit the cockpit and sky. Time had caught up with us. Knocked back in the seat, I said, "My God, I've had it now."[2]

Powers bailed out of the plunging aircraft, parachuted safely to the ground, and was quickly captured by Soviet authorities. He would spend just 17 months in a Soviet prison before being returned to the U.S. as part of an American-Soviet prisoner swap, but the damage had been done. Eisenhower, embarrassed by what had happened, denied any wrongdoing ("nothing that could be considered honestly provocative").[3] But the wily Khrushchev knew he had a trump card, and he played it with customary pomp. On May 5, he made a speech to the wild applause of the Supreme Soviet in the Great Kremlin Palace:

> Comrade Deputies! On the instructions of the Soviet government, I must report to you on aggressive actions against the Soviet Union in the past few weeks by the United States of America...on May Day...an American plane crossed our frontier and continued its flight deep into Soviet territory. The Minister of Defense immediately reported this aggressive act to the government. The government said, "The aggressor knows what he is in for when he intrudes upon foreign territory. Shoot down the plane!" This assignment was fulfilled. The plane was shot down.[3]

The Paris summit later that month was a failure, and quickly U.S.–Soviet relations were staggering badly. The U.S. became obsessed with the communist and pro-Soviet Castro regime in Cuba, and an attempt to overthrow the government failed miserably with the Bay of Pigs debacle in the spring of 1961. In August, East German communists (under the direction of the Soviets) erected the Berlin Wall, dividing the city into two parts impenetrable by land. American military advisers proposed a series of nuclear strikes against the Soviets, citing a U.S. advantage in missiles, which would hold American losses to "only" 3 to 15 million casualties, advice that President John F. Kennedy cautiously turned aside.[4] Kennedy sent more troops into the city and traded tough speeches with Khrushchev, but an actual clash was averted.

As U.S.–Soviet relations spiraled downward, atomic weapons re-emerged into the public eye. There had been a push to resume bomb testing, with American politicians scrambling to take sides. New York Governor Nelson Rockefeller, with his eyes set on national office, had called for a test resumption in the fall of 1959. Kennedy, an all-but-certain presidential candidate, promptly responded by declaring his support for the moratorium.[5] On May 7, 1960, just after Powers' plane crashed, Eisenhower announced that 11 underground nuclear explosions were planned to improve detection methods of enemy activity, but held off the actual tests.[6] Finally, the moratorium broke under the weight of strained relations. Beginning September 1, 1961, the Soviet Union recommenced atmospheric testing, exploding bombs almost daily for the next six weeks. Kennedy had no choice but to match Khrushchev, and the U.S. detonated its first device (underground) in nearly three years in the Nevada desert on September 15,[7] followed by 32 more blasts at the site over the next seven months.[8] Suddenly, environmental radioactivity was in the news again. From September 11 to 17, overall levels in Hartford soared from 1.4 to 65.8 picocuries per cubic meter of air, with many other cities also recording large increases.[9] Levels were highest in the rainy southeast, topped by a reading of 500 in Montgomery, Alabama.[10]

As Soviet and American explosions continued into 1962, concerns rose. A major reason was the soaring level of radioactive chemicals measured in milk, water, and air, the data on which was now much more available to the American public than ever before. An August 14, 1959, directive from Eisenhower mandated periodic monitoring of radioactivity, many of which were taken monthly. By the summer of 1960, about 60 U.S. cities were providing regular measurements available to the public, up from only nine cities reporting between 1957 and 1959; and data were now issued by the U.S. Public Health Service in booklet formats.

With data on dietary radiation more available and with American interest piqued after a decade of Nevada testing, shocking stories began to make the news. In July 1962, reports of iodine-131 levels in Salt Lake City milk between 1500 and 2500 picocuries per liter (normal levels are less than 10) surfaced soon after a flurry of Nevada tests. Although public health officials continued to insist these were well below "safe" levels, 39 milk producers in six counties agreed not to send any fresh milk from cows that had been grazing until iodine levels diminished.[4,11] Mothers across the U.S. began stockpiling powdered milk, rather than feed their children milk that had been contaminated by fresh fallout.[12]

The growing horror of the American public over fallout — a fear perhaps more real than the fear of nuclear war — quickly manifested itself in a series of protests symbolic of the 1960s. Perhaps the first large, organized rally to ban or limit bomb testing occurred on May 19, 1960. On that day, 17,000 gathered in New York's Madison Square Garden heard passionate speeches that included the still-novel slogan "Ban the Bomb." Immediately after, a chanting crowd of 5000 marched from Times Square to the United Nations, led by luminaries such as Socialist leader Norman Thomas and labor titan Walter Reuther. Eleanor Roosevelt co-chaired the rally, and former Republican Presidential candidate Alfred Landon lent it his support.[13] Organized protests began to spread around the country; on Easter Sunday 1962, marchers took to the streets in New York, Philadelphia, Chicago, Boston, and other locations.[14] Protests were almost always nonviolent, increasing support from Americans watching news reports on television.

In addition to sentiment from the general public, scientific information was beginning to accumulate on potential hazards of fallout, applying further pressure on policy makers to halt the testing. In 1958, the same year as Stewart's ground-breaking article on hazards of radiation from X-rays, a United Nations committee made a significant change in the official attitude toward health hazards of radiation. Known as the United Nations Scientific

Committee on the Effects of Atomic Radiation (UNSCEAR), the panel of experts agreed unanimously that bomb test fallout is harmful to human health — specifically by increasing the risk of cancer, leukemia, prenatal conditions, and genetic damage — even in minute amounts. UNSCEAR produced data showing that 1956–57 strontium-90 levels in the northern hemisphere from latitude 25–45 degrees, which includes most of the continental U.S., were about three to four times greater than in the southern hemisphere.[15] All Nevada and Soviet tests had been conducted well north of the equator. Although the debate over just how much harm low levels could cause remained unresolved, a stake had been driven in the heart of the "no immediate danger" credo concerning bomb tests.

The first actual estimates of the number of casualties were also emerging, following up on the initial rough calculations from Pauling and Sakharov. In 1959, Dr. Edward Lewis of the California Institute of Technology computed that between 80 to 1600 U.S. children would develop thyroid cancer in their lifetime due to iodine from bomb test fallout.[16] Even the government began to make estimates, although the projection by the Federal Radiation Council of 40 leukemia and bone cancer deaths a year in the U.S. was extremely conservative.[17] A 1962 article in *Science* magazine by Dr. Ralph Lapp made a strong plea for epidemiological studies on effects of bomb test fallout.[18] Another development pushing toward cessation of bomb tests was the expanded understanding of damage to Hiroshima and Nagasaki survivors. At the beginning of the 1950s, only a few reports of latent cataracts existed; but by the end of the decade, excess numbers of thyroid, lung, stomach, breast, ovarian, and cervical cancer, along with leukemia, had been documented. By 1955, the rate of leukemia for persons within 1000 meters of either bomb was more than five times greater than those over 2000 meters away.[19]

While the pressure was building on Kennedy and his Geneva negotiators to reach an agreement with the Soviets to stop or curtail testing, the President held fast on points such as mutually verifiable inspections of missiles. Kennedy deeply mistrusted Khrushchev and his associates, and sought to earn political points by assuring the American public of their safety from any nuclear attack by keeping the U.S. "Number 1" in the nuclear arms race. Fallout hazards took a back seat, and the testing went on.

However, in October 1962, any inclination to stand fast at the bargaining table on Kennedy or Khrushchev's part was crushed by the frightening near-realization of the worst case scenario of the Cold War.

At about 9 a.m. on Tuesday, October 16, White House aide McGeorge Bundy strode somberly into the President's bedroom with bad news. Just days earlier, U-2 planes had photographed Soviet missile bases in western Cuba, and the CIA was virtually certain that they were designed for offensive, medium-range projectiles easily capable of striking all major U.S. cities with nuclear weapons. Kennedy, publicly maintaining his composure (although aide Theodore Sorenson observed the President to be both surprised by and furious with Khrushchev), proceeded with a number of diplomatic and military maneuvers.[20] All information was kept secret for nearly a week. Finally, on the evening of October 22, a grim Kennedy spoke to the American people to announce the blockade of Cuba by American ships to prevent all incoming offensive weapons:

> Within the past week, unmistakable evidence has established the fact that a series of offensive missile sites is now in preparation on that imprisoned island. The purpose of these bases can be none other than to provide a nuclear strike capability against the Western Hemisphere.
>
> This secret, swift, and extraordinary build-up of Communist missiles...this sudden, clandestine decision to station strategic weapons for the first time outside of Soviet soil, is a deliberately provocative and unjustified change in the status quo which cannot be accepted by this country.[20]

For several days the world held its breath, wondering if Khrushchev and Kennedy could work something out before a war between two nuclear powers broke out. American military forces were prepared to unleash nuclear weapons at the President's command, an option urged by military advisers. On October 25, two Soviet ships were stopped and questioned, and the next day, the American government gave the U.S.S.R. an ultimatum to remove the Cuban missiles within 48 hours. Finally, on October 27, the Soviet leader gave in, agreeing to dismantle the Cuban missile sites (in exchange for the unpublicized pledge by the Americans to abandon nuclear missile bases in Turkey). The world, given a respite from the worst nightmare of the nuclear age, began to move slowly away from that dreadful abyss.

In the aftermath of the near-holocaust raised by the Cuban missile crisis, American and Soviet negotiators moved rapidly to agree on a test ban. The evidence of health hazards from fallout was still growing, but paled in comparison with the compelling power of the near-disaster in Cuba. On July 21, 1963, the U.S., U.S.S.R., and Great Britain announced a tentative agreement had been reached which was signed by Dean Rusk, Andrei Gromyko, and Alec Douglas-Home in Moscow on August 5. The U.S. Senate, despite some protests from a few Republicans (such as Presidential hopeful Barry Goldwater of Arizona) and southern Democrats, ratified the pact by an 80–19 vote; and on October 7, just six weeks before his death, Kennedy signed the Limited Test Ban Treaty. From this point on, no further above-ground tests were conducted by the three powers that signed (the last U.S. above-ground test was on November 4, 1962, on Johnston Island in the Pacific).[8] Although France and China did not sign the treaty, and conducted a small number of atmospheric tests in the late 1960s and 1970s, large-scale testing in the air was over, and the nuclear age closed a chapter in its book.

Soon after the treaty was concluded, U.S. government health officials began to loosen their grasp on fallout data. In August 1963, the Joint Congressional Atomic Energy subcommittee called on a St. Louis citizens group to report on fallout levels it had been compiling. The group contended that 3000 Utah and Nevada children had received excessively high doses of radiation; that dangerous fallout had spread far beyond Utah and Nevada; and that at least seven underground tests had introduced fallout into the atmosphere through "venting."[21] At a news conference, a testy Kennedy reacted to a question about the Utah and Nevada children by saying, "It is a matter of concern...but I would say that as of now, that we do not believe that the health of the children involved has been adversely affected."[22] Interestingly, however, that defensiveness from the White House gave way after the treaty took effect and environmental radioactivity lessened. On October 28, 1964, as Lyndon Johnson campaigned to keep the Presidency, he made the following remarks at the University of New Mexico in Albuquerque:

> We cannot and we will not abandon the test ban treaty to which I just referred, which is the world's insurance policy against polluting the air we breathe and the milk we give our children.
>
> Already that policy has paid off more than you will ever know, and since this agreement was signed and the tests stopped, the dread strontium-89 and iodine-131 have disappeared from the environment. The amount of strontium-90 and cesium-137 has already been, in 1 year, cut in half. This is technical language, but what it means is that we can breathe safely again.[23]

Perhaps the most incriminating evidence showed up in the October 1963 issue of *Radiological Health Data*, the monthly government publication on environmental radiation levels mandated by Eisenhower four years earlier. For the first time, the extent of how much fallout had penetrated the diet of the average American came to light. Some of the data were shocking. For example, the "Smoky" bomb test of August 31, 1957, was not only a large (44,000 tons of TNT) blast, but a "dirty" one in which the fallout was more uncontrolled than the military expected. Radioactive elements with short half-lives quickly made their way into the diet across the country. Table 3.1 shows the levels of several of these elements in raw milk in the five cities covered by the fledgling monitoring program in September and October 1957. Three elements measured each month included iodine-131 (which damages the thyroid gland), strontium-89 (which clumps to the bone and penetrates into the bone marrow, where the immune system is formed), and barium-140 (another bone-seeker). All have half-lives of several days to several weeks.

Table 3.1   Average Picocuries per Liter in Raw Milk, Iodine-131, Strontium-89, and Barium-140 for 5 U.S. cities, September–October 1957

City	I-131 (norm <10)		Sr-89 (norm <5)		Ba-140 (norm <10)	
	Sept	Oct	Sept	Oct	Sept	Oct
New York	300	250	125	190	125	160
Cincinnati	430	300	100	200	125	70
St. Louis	970	890	360	215	530	120
Salt Lake City	740	990	85	150	85	130
Sacramento	10	20	20	35	25	45

The highest levels of each element were recorded in St. Louis, about 1500 miles from the test site, rather than the more proximate Salt Lake City, reflecting the eastward drift of the fallout cloud. Sacramento, north and west (upwind) of the Nevada test site, was spared much of the fallout. All this information was kept quiet by the military, and the news coverage of the test was routine. However, in 1980, the test would resurface because of the public revelation that a number of military personnel had been contaminated with radioactivity when they entered the test site area soon after the blast.

The fallout continued to spread even after the Nevada tests were called off. During the year July 1958–June 1959, strontium-89 in St. Louis milk averaged 250 picocuries per liter (normal between 0 and 10), while Atlanta, 2000 miles from the test site, averaged 110. It became clear that precipitation affected how much radioactivity made its way into the food chain of distant areas; compared to the rainier St. Louis and Atlanta, the drier Salt Lake City averaged 31 picocuries per liter during this time.

Another point that surfaced was the divergent patterns of short-lived and long-lived radionuclides. The longer-lived products tended to collect in the upper atmosphere; settle into the earth's environment only gradually over the next year or so; and decay at a much slower rate than the shorter-lived products. Rather than rocketing up in the weeks after a bomb test, levels of long-lived products rose much more gradually, but stayed in the diet for years, not reaching their peak until the mid-1960s, several years after the Limited Test Ban Treaty went into effect. These patterns became clear after the U.S. Public Health Service's 1968 publication of 1963–66 data.

Table 3.2 shows historical levels of strontium-90 (half-life of 28 years) in New York City's raw milk from 1957 to 1966. New York wasn't the hardest-hit of all American cities, but its patterns of strontium-90 amounts is typical of the nation's. Strontium-90 is a deadly bone-seeking element leading to leukemia, bone cancer, and a variety of cancers and immune disorders. Levels rose during the bomb test periods of the late 1950s, declined after the 1958–61 moratorium, and rose after the resumption of tests in 1961. The highest levels were reached between June 1963 and May 1964, with an average of 27.4 picocuries per liter, compared to a normal rate of under five. Thereafter, strontium's presence declined gradually. One exception is the unusually small amount of 4 picocuries per liter of milk in September 1964.

Table 3.2   Picocuries of Strontium-90 per Liter of Raw Milk
New York City, 1957–66

Month	1957	1958	1959	1960	1961	1962	1963	1964	1965	1966
Jan		6	7	10	8	7	14	27	16	15
Feb		5	7	10	9	7	18	24	18	13
Mar		4	7	11	7	6	11	25	26	14
Apr		4	8	10	7	7	11	25	17	10
May	5	4	12	12	9	15	14	26	15	11
Jun	5	10	14	9	9	16	25	23	19	17
Jul	9	11	15	10	6	15	36	28	20	14
Aug	7	3	12	8	5	19	30	18	13	13
Sep	6	6	4	8	4	17	27	4	14	10
Oct	5	10	8	7	5	19	32	19	11	13
Nov	5	9	9	9	6	17	28	18	16	12
Dec	5	8	9	8	7	14	24	15	14	
Yr Avg	6	7	9	9	7	13	23	21	17	13

Because August had been the first month in New York City history with no recorded rainfall, this sudden drop occurred, but after September the city experienced more typical strontium levels.

Long-lived radioactivity was clearly greatest in areas with high precipitation. For example, in May 1964, when strontium-90 reached an all-time average high of 30.3 picocuries per liter in the U.S., the greatest levels were in the wet southeast, including Little Rock, AR (69); New Orleans, LA (59); Jackson, MS (54); and Chattanooga, TN (52). The other area with high levels was the Dakotas, which were in the direct path of many fallout clouds during the early 1960s; Minot, ND, registered 77 picocuries per liter in May 1964. Conversely, the lowest levels were observed in the dry southwest, including Phoenix, AZ (5); Austin, TX (10); Albuquerque, NM (12); and Las Vegas, NV (13).[24]

In addition to measurements of radioactivity in the diet, the first information on the amounts of radioactivity in the body became available beginning in the late 1950s. In 1958, the Greater St. Louis Citizens' Committee for Nuclear Information organized a drive to collect baby teeth that local six- and seven-year-old children had lost, and analyzed them for strontium-90 content. The initial results were published in *Science* magazine in November 1961, while health effects of atmospheric bomb testing were being debated as part of test ban negotiations. Formula-fed infants born in third quarter 1954 had 3.8 times the concentration of strontium-90 (.589 picocuries per gram of calcium in the tooth) than did those born in the third quarter 1951, the year Nevada tests began (.155).[25] Later reports showed that strontium-90 content in the teeth of St. Louis children born in 1963 soared to levels 15 to 20 times what they were in 1951. Because the buildup of strontium continued through 1965, the bomb testing raised levels of this deadly chemical *30 to 40 times* above its pretesting level, which is close to zero. After atmospheric bomb testing ended, strontium levels in St. Louis fell *in half*, from 1965 to 1969.[26] Since the growing threat was reversed, it remained to be seen whether or not these low-level accumulations had caused health damage to young children, and if so, how extensive the damage was.

The end of above-ground testing was a key event in the story of radiation's effects on human health, but by no means did it end the threat. China and France continued atmospheric testing of a small number of weapons, bringing low levels of fallout to the U.S. from the eastern hemisphere (France conducted its tests in the Pacific, China in its northwest mountains). Sometimes, subsequent fallout levels resembled those of the late 1950s and early 1960s; for example, iodine-131 in Charleston, South Carolina's milk reached 210 picocuries per liter on January 10, 1967, shortly after Chinese test fallout reached the U.S.[27] American underground testing continued as the frenzy of the arms race brought the total stockpile of U.S. atomic weapons to an all-time high of 32,000 in 1967.[28] Explosions beneath the earth's surface brought their own health hazards, especially in the 1960s, as techniques for containing radioactivity were still being ironed out. By January 1964, E.A. Martell of the National Center for Atmospheric Research claimed that much of the increased levels of iodine in U.S. milk in 1961 and 1962 had come from underground testing in Nevada, not from Soviet tests that had been previously blamed by the U.S. government.[29] Martell had been a long-time supporter of the U.S. testing program.

The dangers of underground testing reached their most critical point on December 18, 1970, with the Baneberry explosion at the Nevada test site, 800 feet below the surface. Baneberry opened up a fissure in the rock above the explosion's tunnel, and considerable radioactive dust and gas poured through the opening. Hundreds of AEC employees had to be evacuated, and the fallout cloud, so reminiscent of the 1950s and early 1960s, drifted through Rocky Mountain states such as Idaho, Washington, and Montana.[30] The American testing program moved along at top speed during the 1960s: at the Nevada test site, 429 explosions occurred in the years 1962–70, along with 35 in the Pacific, and 4 in Alaska, Mississippi, and New Mexico. These nine years represent 45% of the 1051 U.S. weapons tests from 1945 to 1992.[8] An estimated 1 of 12 of these explosions leaked radiation into the environment.[31]

Continued Cold War tensions, a vigorous weapons development program, and a military conflict in Vietnam brought nuclear weapon use back into the public debate, only a few years following the Cuban missile crisis. In the 1964 Presidential campaign, Arizona Senator Barry Goldwater startled the nation by stating the "defoliation of the forests by low-yield atomic weapons could well be done."[32] As the war in Vietnam escalated in the 1960s with the expected quick finish having turned into a seemingly endless struggle, more and more support was given to the option of using nuclear weapons. In 1967, Illinois Congressman Paul Findley unsuccessfully pressed the Johnson administration to issue a nuclear ultimatum to the North Vietnamese communists.[33] The 1968 Vice Presidential candidate Curtis LeMay said that if necessary "I would use anything we would dream up, including nuclear weapons."[34] Past and future presidents Eisenhower and Ronald Reagan were not so blatant as LeMay, but publicly refused to rule out support of atomic weapons as an option.

While military developments dominated news coverage of atomic energy during the 1960s, it also proved to be a significant decade for civilian uses of nuclear power. The Price–Anderson Act's limit on liability energized private developers of nuclear power plants, as did considerable government support of plant design and development; by 1962, the AEC had spent about $1.3 billion on reactor research and development, two thirds of the total amount.[35] Electric utilities began to order nuclear reactors in larger numbers. Orders increased from 7 in 1953–57, to 9 in 1958–62, to 70 in 1963–67, to a peak of 97 in 1968–72.[35] By 1966, with only nine plants in operation, the Westinghouse Corporation was already claiming nuclear power to be profitable (Westinghouse, along with General Electric, manufactured most U.S. reactors). Corporate executives from nuclear plant manufacturers spoke boldly about large-scale expansion into all areas of the country, including major cities. John W. Simpson, a Westinghouse vice president, echoed this sentiment:

> It's just a matter of time for such a unit in a major city.
> There's absolutely no reason why it can't be done now.
> I'm afraid those who oppose such plants have axes to
> grind.[36]

This opinion ignored a number of concerns about potential health hazards posed by nuclear power plants. As early as 1956, former AEC official John C. Bugher declared at an American Public Health Association meeting that an atomic power program would present a much greater health threat than nuclear weapons, due to the large quantities of radioactive chemicals emitted into the environment during power generation.[37] The following year, Thomas Parran of the University of Pittsburgh School of Public Health said that the amount of environmental radiation then in existence (much of it bomb test fallout) was "infinitesimal" compared to what would exist under a full-fledged atomic energy program, and expressed worry about health effects as he conceded mankind generally lacked a solid knowledge of radiation's effects.[38] These warnings generally were ignored by industrial and government officials, and these public and private forces backing atomic power's growth moved along snappily with their program.

The expansion of the U.S. nuclear power program coincided with years in which Americans were subjected to shocking and disturbing events that changed their perception of how government worked, and that moved them to more freely challenge authority. Three events, actually chains of events, from 1963 to 1974 stand out. The first began on November 22, 1963, when President Kennedy was assassinated in Dallas. Jolted badly by seeing a young, dynamic leader violently taken from them, the American public received further shocks related to the murder. Two days after Kennedy was killed, his alleged killer was himself gunned down by a man of shady repute; and a 1964 investigation of the assassination headed by Supreme Court Chief Justice Earl Warren was lambasted by a number of researchers and public figures as inadequate, perhaps even a cover-up of a potential conspiracy. One of the most vocal critics was New York attorney Mark Lane. In his book *Rush to Judgment,* Lane blasted the methods and conclusions of the Warren Commission:

> When principles of law and rules of evidence are dispensed with or relaxed, reason must continue to prevail. The Commission disregarded these rules and principles, making no explanation and adopting no substitute. Hearsay evidence was freely admitted, while crucial eyewitness testimony was excluded. Opinions were sought and solemnly published, while important facts were rejected, distorted, or ignored. Dubious scientific tests were said to have proved that which no authentic test could do. Friendly witnesses gave testimony without fear of criticism or cross-examination, were led through their paces by lawyers who, as the record shows, helped to prepare their testimony in advance and were asked leading questions; while those few who challenged the Government's case were often harassed and transformed for the time being into defendants. Important witnesses with invaluable information to give were never called, and the secrecy which prevailed at the hearings was extended, in respect to many important details, for another 75 years.[39]

Coming right on the heels of Kennedy's death, the Vietnam War rattled Americans even further. Lyndon Johnson succeeded in manipulating two hostile encounters with North Vietnamese troops in the Gulf of Tonkin in August 1964 into a congressional resolution that served as the equivalent of a war declaration. Two years later, as thousands of American combat troops poured into Southeast Asia, Congress held hearings charging that Johnson's administration had defrauded the public in 1964, to no avail. Escalation continued, and American casualties mounted. The massive North Vietnamese Tet offensive of January 1968 showed the public they had been misled by Johnson into thinking that the enemy was nearly defeated. Public disapproval turned into a growing number of demonstrations which forced Johnson's departure from the 1968 Presidential race and aided the subsequent defeat of Vice President Hubert Humphrey by Richard Nixon later that year. The eventual withdrawal of U.S. troops from Vietnam and the 1975 takeover by communists further divided and embittered Americans about the Vietnam experience.

The third notable shock of those years started with the arrest of five men breaking into Democratic party national headquarters in 1972. Within days, actions were taken in the Nixon White House to thwart any subsequent investigation of the crime. The affair remained out of the limelight until the following year, when investigations by Congress and an independent prosecutor revealed a stunning network of covert activities and cover-up. Nixon was bombarded by continued questions from the press, and his protests of innocence were becoming more desperate as he sank lower and lower in the polls. On November 17, 1973, he snapped to a press conference in Orlando, Florida:

> In all my years of public life, I have never obstructed justice. And I think, too, that I could say that in my years of public life, that I welcome this kind of examination, because people have got to know whether or not their President is a crook. Well, I am not a crook.[40]

Overwhelmed by public outrage and facing possible impeachment in Congress, Nixon was hounded into resignation in August 1974, and although he received a pardon from successor Gerald Ford and was never prosecuted, a number of his underlings were convicted of various crimes.

While a number of other events opened the public's eyes in those years, the Kennedy assassination, Vietnam, and Watergate played a special role. More than any other historical transpirations of that era, they served to shock Americans out of a slumberous public trust, remind them that authority was not to be automatically accepted, and reinforce the American right to question and protest. In addition, the Vietnam experience made Americans question how far the nation should go in fighting the Cold War against Communism. Instead of just a "fad" reaction that faded away, these newly found beliefs held fast in years to come. No institution was spared from these attitudes, and the nuclear industry soon ran headlong into it, putting it on the defensive for the first time since the Atomic Age began in the 1940s.

Similar to the experience with atomic weapons tests, the first health-related fears on nuclear power plants originated from the general population, not from public officials. It didn't take long for citizens' concerns to be converted into organized action against nuclear plant development. The first such action surrounded the proposal of New York's Consolidated Edison to have Westinghouse build a large, 1,000,000 kilowatt reactor and place it in Ravenswood, Queens. The site lay just across the East River from Manhattan, the heart of New York City, meaning that over 5,000,000 people would be living within five miles of the reactor. In the latter part of 1963, organized groups picketed at the site and sent petitions to President Kennedy and New York Mayor Robert Wagner. On August 3, former AEC chairman David

Lilienthal spoke out against the plant, and against any nuclear facility located near large populations because of the safety threat:

> We must think about a great proliferation of these plants all over the country. The pressure of engineering convenience and costs will bring down these plants more and more into densely populated areas. To start down that road without first completely licking the problem of risks, or dependability in a regionwide system, is a foolhardy course.[41]

Despite Westinghouse and Con Edison's repeated assurances of the safety of the plant, plans were scrapped in January 1964. The Ravenswood experience proved to be only the first of many confrontations between the nuclear industry and the public over health issues. In 1968, the first organized public forum on health effects of nuclear power was held at Stratton Mountain, Vermont, not far from the proposed Vermont Yankee nuclear power plant in Vernon, which had evoked considerable opposition. The AEC caused a flap when it refused to send any representatives to the conference.[42] By 1969, buckling to pressure by George D. Aiken, one of the Senate's most senior and powerful members, the AEC sent 39 staff members to meet with the public at an open meeting in Burlington, Vermont.[43] The AEC testimony, some of which was given by Chairman Glenn Seaborg, did little to persuade antinuclear groups of plant safety. In 1970, a last-ditch appeal was made for the state to block the opening of the plant. The case for adverse health effects was presented in blunt terms: John Gofman of the Lawrence Radiation Laboratory in California testified against the plant, asserting that the permissible annual AEC level of 170 millirems of radioactive substances in the environment caused 32,000 cancer deaths each year.[44] Gofman and his associate Arthur Tamplin had first made this startling assertion the previous year.[45] Despite these pleas, Vermont Yankee began operations in 1972.

Elected officials, seeing the public anxiety over the health effects of radiation, began to openly question nuclear power development. In 1968, Senator Edward M. Kennedy suggested a moratorium on licensing more power plants until a thorough study on potential dangers to the public could be made. Three years later, Senator Mike Gravel submitted a similar proposal. The AEC was becoming more detached from the public and elected officials and was viewed with greater hostility. In 1969, New York Congressman Jonathan Bingham proposed legislation that would have transferred regulatory authority for licensing power plants away from the AEC to the Department of Health, Education, and Welfare.

Worries over plants adjoining big cities became especially acute. In 1965, the AEC was presented with applications for new large reactors at Indian Point (close to New York City) and Dresden (near Chicago). The AEC's safety committee debated whether to write a letter of warning to the commission because of the potential safety hazards near large populations; AEC official Harold Price quashed the suggestion, warning of "a major public relations problem" if requests for reactors near big cities were denied.[46] Both the Indian Point and Dresden proposals were approved and the reactors were built. Despite mounting worries over the safety of civilian nuclear reactors, the AEC inexplicably responded by *raising*, not tightening, the permissible levels of exposure near nuclear plants. The AEC, worried that any perceived or actual threat might hamper the weapons testing program, was taking no chances.

As concerns in many localities about the health effects of radiation sprang up, much of the discussion centered on the potential of a nuclear accident, or the harm caused to aquatic life from the release of hot water from plants. Very little focus was placed on effects of exposure by local residents to routine emissions of low doses of radiation over many years. The public was obviously more frightened by the specter of a nuclear accident, and focused its attention accordingly. The scientific community, populated by many atomic scientists with ties to either the military or civilian development of the atom, believed effects of such low levels of routine emissions to be so small as to not pose a threat. They had observed effects on Hiroshima and Nagasaki survivors, as well as inhabitants of the Marshall Islands during the years of Pacific bomb tests. Although cancers and genetic disease caused by these events were significant, scientists assumed that reducing doses would accordingly reduce excess cases of affected people. For example, a dose one-millionth of the Hiroshima bomb would be expected to cause one-millionth of the excess cancer cases in a population of similar size.

This "linear dose response curve" went unchallenged for many years, although questions were raised (beginning with Stewart in 1956 and MacMahon in 1962) on the potency of low doses. In 1969, Ernest Sternglass of the Department of Radiology at the University of Pittsburgh became embroiled in a controversy over trends in childhood leukemia and stillbirths in the area comprising the three adjacent upstate New York cities of Albany, Schenectady, and Troy. The area had been hit by a violent rainstorm in April 1953, which brought some of the radioactive cloud from the Simon bomb test in Nevada into the food chain (see Chapter 2). While the increase in radioactivity was considerable, levels were well below those of Hiroshima, Nagasaki, and the Marshall Islands. Sternglass publicized data showing that leukemia deaths for local children under 10 jumped from 9 to 30 between 1952–55 and 1959–62.[30] He argued that this increase, beginning about five years after exposure, matched the same pattern observed by Stewart for children exposed to X-rays *in utero*. He also presented information that the fetal death rate in New York state had risen during the bomb test years, when strontium-90 levels in the environment were rising.[47] Although the New York State Health Department issued data refuting Sternglass and strongly disputing his claims, the public had been exposed to a disturbing scenario.

Perhaps Sternglass' greatest controversy was caused by an article published in the September 1969 issue of *Esquire* magazine. In May of that year, Sternglass had roused an assembly of scientists at the 9th Hanford Biology Symposium into a heated debate by proclaiming that between 1950 and 1965, about 375,000 additional infants had died before their first birthday due to atmospheric bomb testing.[48] Nothing was reported in the media in the few months after the Hanford meeting, but that changed when *Esquire* went to press. The stirring article, actually a special insert ordered at the last minute by editor Harold Hayes entitled, "The Death of All Children," also included data on childhood cancer and fetal deaths.[49]

Sternglass was immediately pilloried by the scientific establishment, and the AEC enlisted Arthur Tamplin of the University of California to produce a study rebuking Sternglass. Tamplin's calculated excess infant mortality of about 4000 was much lower than Sternglass', but proved to be still too high for the powers at the AEC, engaged in carrying out its Cold War program. A dispute between Tamplin and his partner John Gofman broke out; Gofman resigned his position as Associate Director of Livermore Laboratories at the University of California, and the AEC removed 11 of Tamplin's 12 research associates, seriously impairing his research efforts.[30] The clash over infant mortality in 1969–70 left many questions unanswered, but pushed the debate over health effects of low-level radiation exposure to a new plane.

The early 1970s may be seen as the high water mark of the American nuclear program. The vigorous buildup of atomic weapons had brought the number to 29,000 in 1974, down just slightly from the peak of 32,000 in 1967.[50] The nuclear power program was flourishing as plants continued to open and utilities ordered more. By 1974, a total of 242 reactors had been ordered (nearly 60% between 1970 and 1974),[35] and 40 of these were already operating and producing 5% of the country's electricity. The Nixon administration boldly set a goal of 1000 reactors in operation by the year 2000.[35] Not a single ordered plant had been cancelled by 1972. The nuclear industry had not incurred any serious accidents (at least none publicly known). But this was all to change, and the nuclear industry's fortunes began to sag. Much of the turnaround had to do with the health and safety effects of radiation, a subject still in relative infancy in the early 1970s. And once again, it was the general public that galvanized the movement that turned the nuclear story around.

In the early 1970s, only a scant few voices were raised about the dangers of low-level radiation exposure. Sternglass was the most vocal and most recognized. In 1971, he informed a public hearing in New York that infant mortality had increased near the Indian Point and Brookhaven reactors, both in the New York metropolitan area, because of low-level emissions.[51] In 1973, he released data showing cancer had increased 30% and leukemia 70% near Shippingport, Pennsylvania, in the 10 years after the reactor opened in 1957.[52] Others, such as Alice Stewart, openly questioned whether there was ever a chance of safe levels of human exposure to radioactivity. The AEC tried to stem the ripple of unrest by proposing lower exposure standards for citizens in 1971. The proposal called for no more than 5% of ordinary background radiation each year.[53] However, the AEC failed to consider that fission products such as strontium and iodine do not exist in nature, making background radiation a suspect method of setting "safe" levels. The AEC was beginning to come under fire in the early 1970s for its lenient policies in penalizing unsafe nuclear power plants.[54] The dual AEC role of protecting public safety and promoting nuclear power was seen as a conflict of interest, and in 1974, during a period of considerable legislative activity to protect the environment, Congress passed a law abolishing the Agency, and placed authority for health and safety issues (but not nuclear power promotion) in the hands of the newly formed Nuclear Regulatory Commission.

One event of the early 1970s that went unnoticed at the time but later came to play a major role in enhancing knowledge of radiation's health effects was a discovery by Canadian physician and biophysicist Abram Petkau. Working as a researcher in Manitoba, Petkau tested an idea in 1971 that got startling results. Adding a very small amount of radioactive sodium-22 to beef brains immersed in water caused the membranes of the brain cells to break much faster than he previously expected. Repeating his experiment again and again with the same results, Petkau concluded that the longer and more drawn-out the exposure to radioactivity, the smaller the dose needed to damage the cell.[55]

Even though he published his findings in the journal *Health Physics* in 1972, Petkau received little recognition for his discovery. The eventual implications, however, were enormous. Petkau's work made it possible to understand the difference in health effects between short, high-dose bursts of radiation (such as an X-ray or immediate effects of an atomic bomb), and prolonged, low-dose exposures such as living near a nuclear facility that steadily releases small amounts of radioactivity over many years. First, the X-ray is a type of gamma ray focused on one particular part of the body, such as the skull or chest; nuclear plant emissions are a cocktail of different alpha, beta, and gamma ray emitters that damage many organs of the body (iodine seeks the thyroid gland, strontium clumps to the bone, cesium inserts itself into soft tissues, etc.). Second, the short, high-dose burst from an X-ray or an atomic bomb kills cells; but by acting in the way Petkau demonstrated, prolonged small doses are more likely to break the cell membrane and leave the cell impaired than to kill the cell outright. An impaired cell may be able to repair itself, but it also may continue to function abnormally; and it may not be able to reproduce properly, which could lead to the development of cancer.

Sternglass quickly recognized the value of Petkau's experiment, and adopted the belief that low doses of radioactivity were more harmful *per dose* than high doses, especially short bursts like X-rays and the initial explosive action of atomic bombs. Sternglass thus supported a "supralinear dose response curve" between radiation exposure and health effects, i.e., more cancer cases per rad at lower doses, whereas most other scientists, even a pioneer like Stewart, believed in the "linear dose response curve," i.e., 1 rad produces 1 cancer case, 2 rads produce 2 cancer cases, etc., no matter what the dose. Sternglass asserted that basing any estimates of cancers from low-dose nuclear reactor emissions on the experience of Hiroshima and Nagasaki survivors was erroneous. This stance would make him more controversial than ever, but the debate over low-level radiation's effects would also not go away as the years went on and the adverse health effects mounted.

Aside from the data Sternglass had been collecting, not much statistical evidence had linked adverse health effects to radiation. In the mid-1970s, however, landmark research by Alice Stewart, Thomas Mancuso, and George Kneale established a correlation between cancer risk and radiation doses among workers at the Hanford plant. Specifically, workers with 16 or more years of experience at Hanford had twice the risk of dying of cancer compared to non-atomic workers of the same age living in the region.[56] The results were published in *Health Physics* in 1977, but not before the federal government worked to stop the publication. Subsequently, Mancuso lost federal funding to continue the project, even though Washington state epidemiologist Samuel Milham produced an independent study showing elevated cancer rates among Hanford workers. The government also made an unsuccessful attempt to confiscate the study data.[57]

While the 1970s represent the apex of the American nuclear power industry, they also marked the start of its slide. Public interest groups such as Ralph Nader's Public Citizen repeatedly issued watchdog reports of the dangers of nuclear power plants, and became involved in litigation as well. Other, more "establishment" groups sometimes joined in; in 1973, the Rand Corporation, a large defense contractor, questioned whether nuclear power was the best source of energy for the America of the future.[58] The nuclear industry found itself increasingly on the defensive about safety and health concerns. Accordingly, more care was placed in the production and assembly of nuclear reactors, increasing the time required for startup, and increasing bottom-line costs. Realizing that the concept of cheaper, cleaner energy was slipping through their fingers, the utilities pulled an about-face. In 1972, the first cancellation of a reactor in history occurred, after 177 orders,[35] and in 1974, Consumers Power Company of Michigan cancelled two 1,150 megawatt reactors, the first cancellation of large reactors in an advanced stage of planning.[59] Between 1975 and 1982, 81 reactors were cancelled, compared to only 11 ordered.[35] While previously ordered reactors continued to begin operations, the nuclear industry's fortunes were clearly fading, due largely to the health and safety issue.

Public discontent began to boil over. Even though there had been protests in the 1960s against nuclear power, the situation near the proposed Seabrook site in New Hampshire eclipsed all previous activity. On April 30, 1977, 2000 demonstrators shocked the nation when they moved onto the construction site, vowing to stay until physically removed or the operators cancelled the plant.[60] The following year, 20,000 more descended on Seabrook in a massive rally that heightened the public debate, and protests marred Seabrook's history until it finally opened in 1990. A number of other protests took place in 1977 and 1978 at nuclear sites, including the Rocky Flats, Colorado, plant where plutonium was made for atomic bombs.

Again, much of the worry over nuclear power was the potential for accidents, and the mid-1970s actually produced two of them, neither of which was catastrophic, but which was a foreboding of things to come. On March 22, 1975, at the Brown's Ferry plant in northern Alabama, a technician used a candle to see better in a dimly lit part of the reactor and started a fire. The blaze spread, lasting over 7 hours and damaging cables and the emergency cooling system, although authorities claimed little radioactivity was released. On September 24, 1977, at the Davis–Besse plant in Ohio, a pressure relief valve became stuck in an open position, causing the vital supply of cooling water to flow out of the reactor. Luckily, because the plant was running at just 9% of capacity and the problem was detected in just 20 minutes, a minimum of radioactivity escaped from the plant. The Davis–Besse reactor was designed and built by the Babcock and Wilcox company, which had designed a number of other plants, including the as-yet-unknown Three Mile Island complex near Harrisburg, Pennsylvania.[46]

Spurred by the public concerns of a potential major accident plus the mishaps at Brown's Ferry and Davis–Besse, some scientists began to speak out. In 1976, an engineer at Indian Point resigned in a huff, proclaiming the plant's number 2 reactor to be "an accident waiting to happen."[61] Later that year, three engineers who had recently resigned from General Electric told the Congressmen on the Joint Committee on Atomic Energy

> ...the cumulative effect of all design defects and defi-
> ciencies in the design, construction, and operation of
> nuclear power plants makes a nuclear power plant
> accident, in our opinion, a certain event. The only ques-
> tion is when and where.[35]

References

1. Cagin, S. and Dray, P., *We Are Not Afraid*, Bantam Books, New York, 1989, 68-75.
2. Powers, F. G., *Operation Overflight: The U-2 Spy Tells His Story for the First Time*, Holt, Rinehart, and Winston, New York, 1970, 82.
3. Bechschloss, M., *May Day: Eisenhower, Khrushchev, and the U-2 Affair*, Harper and Row, New York, 1986, 43, 265.
4. Miller, R., *Under the Cloud: The Decades of Nuclear Testing*, The Free Press, New York, 1986, 335.
5. *The New York Times*, November 3, 1959, 3.
6. *The New York Times*, July 18, 1960, 27.
7. *The New York Times*, September 16, 1961, 1.
8. Norris, R. and Cochran, T., *United States Nuclear Tests, July 1945 to 31 December 1992*, Nuclear Resources Defense Council, Washington, DC, 1994, 32-3.
9. *The New York Times*, September 19, 1961, 4.

10. *The New York Times*, September 20, 1961, 6.
11. *The New York Times*, August 2, 1962, 9.
12. *The New York Times*, March 17, 1962, 8.
13. *The New York Times*, May 20, 1960, 1.
14. *The New York Times*, April 22, 1962, 3.
15. *The New York Times*, August 11, 1958, 1.
16. *The New York Times*, June 21, 1959, 76.
17. *The New York Times*, June 2, 1962, 1.
18. Lapp, R. E., Nevada fallout and radioiodine in milk, *Science*, September 7, 1962, 756-7.
19. *The New York Times*, March 17, 1962, 8.
20. Sorenson, T., *Kennedy*, Bantam Books, New York, 1966, 759, 793.
21. *The New York Times*, August 22, 1963, 12.
22. *The New York Times*, August 21, 1963, 14.
23. U.S. Government Printing Office, *Public Papers of the Presidents of the United States: Lyndon Johnson, 1963–64*, Washington, 1965, 1490.
24. U.S. Public Health Service, *Radiological Health Data*, Washington, DC, September 1964, 423.
25. Reiss, L. Z., Strontium-90 absorption by deciduous teeth, *Science*, November 24, 1961, 1669-73.
26. Rosenthal, H., Accumulation of environmental 90-Sr in teeth of children, in *Proceedings of the Ninth Hanford Biology Symposium*, Hanford, WA, 1969, 163-71.
27. *The New York Times*, January 21, 1967, 33.
28. Udall, S., *The Myths of August*, Pantheon Books, New York, 1994, 143.
29. *The New York Times*, January 12, 1964, 40.
30. Sternglass, E., *Secret Fallout: Low-Level Radiation from Hiroshima to Three Mile Island*, McGraw-Hill, New York, 1981, 180.
31. *The New York Times*, August 21, 1970, 53.
32. Lokos, L., *Hysteria 1964: The Fear Campaign Against Barry Goldwater*, Arlington House, New Rochelle, NY, 1967, 106.
33. *The New York Times*, May 5, 1967, 1.
34. *The New York Times*, October 4, 1968, 1.
35. Hertsgaard, M., *Nuclear Inc.: The Men and Money Behind Nuclear Energy*, Pantheon Books, New York, 1983, 41.
36. *The New York Times*, April 10, 1966, III 1.
37. *The New York Times*, November 15, 1956, 1.
38. *The New York Times*, May 28, 1957, 15.
39. Lane, M., *Rush to Judgment*, Fawcett Crest, New York, 1967, 339.
40. Woodward, B. and Bernstein, C., *The Final Days*, Simon and Schuster, New York, 1976, 94.
41. Lilienthal, D. E., *The Journals of David E. Lilienthal: Volume 5: The Harvest Years, 1959–63*, Harper and Row, New York, 1971, 497.
42. *The New York Times*, September 15, 1968, 61.
43. *The New York Times*, September 12, 1969, 15.
44. *The New York Times*, September 13, 1970, 35.
45. Gofman, J. and Tamplin, A., *Population Control Through Pollution*, Nelson Hall Company, Chicago, 1970.
46. Ford, D., *The Cult of the Atom: The Secret Papers of the Atomic Energy Commission*, Simon and Schuster, New York, 1982, 88, 92.
47. *The New York Times*, October 12, 1969, 50.

48. Sternglass, E., Evidence for low-level radiation effects on the human embryo and fetus, in *Proceedings of the Ninth Hanford Biology Symposium*, Hanford, WA, 1969, 693-717.

49. Sternglass, E., The death of all children, *Esquire*, September 1969, 1a-1d.

50. *The New York Times*, September 26, 1993, IV 3.

51. *The New York Times*, May 12, 1971, 24.

52. Sternglass, E., *Radioactive Waste Discharges from Shippingport Nuclear Power Station and Changes in Cancer Mortality*, University of Pittsburgh Department of Radiology, 1973.

53. *The New York Times*, June 18, 1971, 1.

54. *The New York Times*, August 25, 1974, 1; November 10, 1974, 1.

55. Gould, J. and Goldman, B., *Deadly Deceit: Low-Level Radiation, High-Level Cover-Up*, Four Walls Eight Windows, New York, 1990, 173.

56. Mancuso, T., Stewart, A., and Kneale, G., Radiation exposure of Hanford workers dying of cancer and other causes, *Health Physics*, 1977, 369-85.

57. Bertell, R., *No Immediate Danger: Prognosis for a Radioactive Earth*, The Book Publishing Company, Summertown, TN, 1985, 88-91.

58. *The New York Times*, April 8, 1973, IV 14.

59. *The New York Times*, June 29, 1974, 1.

60. *The New York Times*, May 1, 1977, 26.

61. *The New York Times*, February 10, 1976, 1.

chapter four

The '80s and '90s: disasters, reflection, and revisionism

The course of U.S. nuclear history changed, drastically and permanently, on an early spring day near Harrisburg, Pennsylvania.

> In the early morning hours of Wednesday, March 28, 1979, Three Mile Island Unit 2 was functioning normally, under full automatic control. No one on duty at the plant was a qualified nuclear engineer, or even a college graduate, nor had anyone there received detailed technical training on how to handle complex reactor emergencies.
>
> The maintenance workers, who were not required to have federal licenses or to go through any federally approved training programs, were trying to cope with a problem common to every type of plumbing system — a clogged pipe. The pipe in question was a small one coming from one of eight tanks, known as polishers, that removed impurities from the main feedwater system.
>
> Shortly before 4 a.m., Frederick Scheimann, the Unit 2 shift foreman, went to the basement to see how work on the pipe was progressing. He discussed the problem with (technician Donald) Miller and with Harold Farst, the other technician on duty. Scheimann climbed up on top of a larger pipe so that he could look into the polisher through a glass window. "All of a sudden, I started hearing loud, thunderous noises, like a couple of freight trains," he said later. He jumped down from the pipe, heard the words "Turbine trip, turbine trip" over a loudspeaker, and rushed to the control room. The maintenance crew working on the

> polisher had accidentally choked off the flow in the
> main feedwater system, forcing Unit 2's generating
> equipment — its turbine and reactor, which had been
> operating at ninety-seven percent of full power — to
> shut down. The equipment was suddenly tripped at
> thirty seven seconds past 4 a.m.[1]

This halting of the flow of cooling water began a nightmarish chain of events. For 16 hours, the core of the Unit 2 reactor had inadequate cooling water, creating the danger of a meltdown of the hot, highly radioactive core. A number of fuel rods, also containing great amounts of radioactivity, were damaged. Worse, radioactive gases collected in an auxiliary water tank, creating the possibility of a serious explosion. A state of emergency was declared at 7:24 that morning, barely three hours after the initial problem. In the early morning of March 30, two days later, plant operators deliberately released radioactive gas into the atmosphere, in a sort of "burping" action, when the dangerous gas was transferred into a safe storage tank. Over two hours later, the state Emergency Management Agency, acting on Governor Richard Thornburgh's orders, suggested persons living within 20 miles of the reactor be evacuated, with special appeal for all pregnant women and small children to leave the area. By April 3, the gaseous bubble of hydrogen was eliminated, taking with it any imminent danger of an explosion, but not until 147,000 panic-stricken Pennsylvanians had fled the region.[1]

The impact on the public was immediate and devastating. The news media descended on Harrisburg en masse, providing Americans with many frightening stories about the damaged reactor. The recently released movie "The China Syndrome," starring Jack Lemmon and Jane Fonda and describing a nuclear meltdown, saw an upsurge at the box office. Within five weeks, a large rally had been organized and 65,000 protestors demonstrated at the Capitol in Washington against the proliferation of nuclear power.[2]

Officials tried to put their best face on the situation. President Jimmy Carter and his wife Rosalynn even visited the plant during the emergency to assure the nation that the situation was safe. Carter promised immediate action, and on April 11, ordered a thorough investigation into the accident. However, Carter's order forbade reviewers from assessing any nuclear reactor other than Three Mile Island.[1]

How bad, exactly, was the accident? Sternglass flew immediately to Harrisburg with a survey meter that measured radioactivity levels, determined to find his own answers. Even before his plane landed, Sternglass noted a rise in radioactivity to as much as 15 times the usual level. On land, levels were consistently high, but unevenly distributed, indicating that Three Mile Island's releases had deposited in "hot spots."[3] Later that year, when the state prepared a report for the Presidential commission, Pennsylvania Health Commissioner Gordon MacLeod became entangled in a dispute with other state officials over how serious the situation had been. MacLeod, who felt the report understated the actual hazards, was soon dismissed by Governor Thornburgh.[3]

The calculation of how much radioactivity eventually escaped into the environment from Three Mile Island was put at 14.2 curies, or 14.2 trillion picocuries of iodine-131 and longer-lived chemicals, by the Nuclear Regulatory Commission. This figure was higher than all other U.S. reactors for 1979, but not by all that much. For example, the Oyster Creek 1 reactor in New Jersey emitted 9.3 curies that year, while Dresden 2 and 3 in Illinois released 7.0 curies.[4] Furthermore, the accuracy of the 14.2 curie figure may be questioned in light of Sternglass' measurements and public skepticism over government veracity; one week after Three Mile Island, 68% of Philadelphia residents (downwind of the plant, about 100 miles away) believed that the peril was greater than that stated by the government.[5]

The assessment of health effects from the accident began almost immediately and continues nearly 20 years after the accident. In the counties downwind of the reactor (under 100 miles), the number of babies born with hypothyroidism jumped from 9 to 20 in the nine months after the accident, while cases in upwind counties fell from 8 to 7.[6] Hypothyroidism, an underactive thyroid gland due to lack of hormone production, becomes more prevalent in populations exposed to radioactive iodine, released in large quantities at Three Mile Island. In Lancaster County, the closest area downwind of the reactor, 6 babies were born with the condition, whereas just over one case would be normally expected.[7]

Another immediate effect of Three Mile Island on the especially vulnerable fetuses and newborns showed up in deaths to babies less than one year old. In Dauphin County, where Three Mile Island is located, death rates in the first month and first year of life soared by 53.7% and 27.5%, respectively, between 1978 and 1979. The five closest downwind counties, all within 50 miles to the east of the reactor, also saw substantial rises in both measures in 1979 (5.8% and 6.0%). Conversely, infant mortality fell in the rest of the state and the United States as a whole. Sternglass showed that the steepest 1979 rises in Pennsylvania infant mortality occurred from April to December, the nine months following the accident.[3]

Longer-term effects were the next to be addressed. In 1985, the Pennsylvania Health Department released a study showing no excess cancer in the five years after Three Mile Island.[8] However, this analysis may have been hampered by its failure to examine different types of cancer, especially those most sensitive to radiation. The 1990 book *Deadly Deceit* by Jay Gould and Benjamin Goldman presented a much more precise analysis. Studying the ten counties closest to Three Mile Island (total 1979 population of about 1,500,000), similar to the Health Department, Gould and Goldman found no change in deaths from all causes, heart disease, and all cancers combined in the five years after the accident. However, they found childhood cancers, other infant diseases, and deaths from birth defects were *15% to 35% higher* than they were before the accident; the earlier low rates had all reached

Table 4.1 Changes in Mortality Rates in 10 Counties
Nearest Three Mile Island, 1968–73 vs. 1979–83

| Cause of Death | % Above/Below PA Rate | | % Change |
	1968–73	1979–83	
All causes	+2	+1	−1
Heart disease	+7	+5	−2
All cancers	−1	0	−1
Lung cancers	−15	−15	0
Infant diseases	−12	+1	+13
Infant birth defects	−11	+8	+19
Breast cancers	+4	+11	+7
Child cancers	−23	+4	+27

levels well above the national average. Breast cancer death rates were also somewhat elevated (up 7%). Moreover, increases in these 10 counties far exceeded those elsewhere in Pennsylvania (Table 4.1). The two researchers conclude that the most vulnerable populations — fetuses and infants living nearest to the plant — appear to have been harmed from Three Mile Island; and women were victimized by the near-meltdown through increased breast cancer, a disease known to be sensitive to radiation from studies of Hiroshima and Nagasaki survivors.

Even more incriminating evidence was uncovered by the National Cancer Institute in its 1990 analysis of cancer mortality near nuclear power facilities. The NCI looked only at the three closest counties to Three Mile Island (Dauphin, Lancaster, and York, with nearly 1 million inhabitants in 1979), and found startling mortality jumps in several types of radiation-related cancers just one to five years after the accident. Table 4.2 shows how 1980–84 death rates in these three counties were considerably higher than 1970–74 (before the plant opened) for leukemia, female breast, thyroid, and bone and joint cancers. The change in deaths from cancer among children under 10 was especially dramatic, nearly doubling compared to the national rate. While these data are available to the public in the NCI report, the federal government did not deem these results to be significant, and made no mention of them in its summary of the report.

Three Mile Island was not the first instance in which the American population had been exposed to potentially harmful doses of radiation, nor may it have been the worst exposure. Specifically, 11 years of living downwind from 100 above-ground bomb tests at the Nevada site may have placed Utah and Nevada residents in greater jeopardy. However, Three Mile Island became the most studied population of the nuclear age, a sign that greater concern over health effects of radiation was permeating the science world and that analytical methods of assessing harm to the population were improving. The accident helped draw attention to elected officials, as more joined the ranks of nuclear skeptics or opponents after 1979. Even utilities that operate nuclear plants may have been swayed; no nuclear reactors have been ordered in the U.S. since 1978.

Table 4.2 Changes in Death Rates, Selected Cancers in Dauphin, Lancaster, and York (PA) Counties, 1970–74 vs. 1980–84

Cause of Death	Deaths		% Above/Below U.S.		% Change
	1970–74	1980–84	1970–74	1980–84	
All Ages					
All cancers except leukemia	6972	8628	+0	+0	+0
Leukemia and aleukemia	283	374	–9	+6	+15
Female breast cancer	707	893	+6	+13	+7
Thyroid cancer	17	22	–22	+8	+30
Bone and joint cancer	40	26	+5	+14	+9
Age 0–10					
All cancers except leukemia	15	22	–35	+26	+61
Leukemia and aleukemia	15	14	–30	+33	+63

Data on health effects of the accident lent credence to the growing belief that man-made causes were behind much of the rising cancer rates besieging American society. This principle had been first proposed by Rachel Carson in her 1962 epic *Silent Spring* and expanded on by Samuel Epstein of the University of Chicago.[9] Epstein's 1979 book, *The Politics of Cancer*, presented voluminous evidence of cancer limited to a variety of man-made sources. Although not mentioning ionizing radiation specifically, Epstein nevertheless invoked the principle of synergy; that is, two or more carcinogens, working together, are more toxic than the sum of each working alone.[10] This principle would be invoked more and more as the study of radiation exposure's effects matured.

The accident at Three Mile Island occurred at an odd time. Just a year later, Ronald Reagan won a landslide election to the Presidency, while Republicans gained control of the Senate for the first time in almost 30 years. Reagan and his adherents charged into office with a clear mandate to carry out positions such as a strong anti-communist militarism, and laissez-faire attitude toward government in business (including the nuclear industry), and reduced regulations (such as in the environmental area). Among Reagan's efforts were the acceleration of defense spending and arms systems, including a greater nuclear arsenal and more weapons testing beneath the Nevada ground. Arms reduction talks with the Soviets had resulted in several treaties beginning in 1972. However, Reagan's insistence on enacting the "zero option" proposal to cancel American development of Pershing missiles in exchange for the dismantling of intermediate-range Soviet missiles stalled negotiations, even though the proposal eventually fizzled out.[11] Cold War tensions between the two powers resurfaced, perhaps best epitomized by Reagan's 1983 denunciation of the U.S.S.R. as an "evil empire." In addition, the Reagan administration sought to revive the lagging nuclear power industry. Spending during Reagan's first year was cut for every domestic program except nuclear power development. In October 1981,

Reagan proposed streamlining the licensing period of nuclear reactors from 10 to 14 years to 6 to 8 years, in addition to calling for swift construction of a nuclear waste repository and reprocessing of waste into weapons-grade plutonium.[12]

Despite the efforts of Reagan and the Republicans, greater revelations of radiation's health effects had begun to convert nuclear progress into decline; along with growing realization that nuclear power was not as inexpensive as once believed, the health effects issue would prove to overpower conservative interests. Three Mile Island was the most obvious focus of the new developments — measuring adverse health effects of a nuclear power plant accident on the local population — but there were others. Low-level waste amassing at nuclear dumps was one emerging specter. Before 1970, most of this waste (items such as contaminated clothing, not to be confused with high-level waste, or used fuel rods) was put into metal barrels and casually dumped into the ocean. In time, the barrels began to corrode and leak, and by 1980, 25% of them were leaking radioactive chemicals into the deep seas. Authorities stopped ocean dumping in 1969, and began to dispose of waste in long underground trenches at various sites. By the late 1970s, this method also proved potentially hazardous, as leakage caused more contamination problems. In the late 1970s, a number of sites were closed either temporarily or permanently, including West Valley, NY, Morehead, KY, Beatty, NV, Barnwell, SC, and Hanford, WA.[13]

High-level waste also posed problems. In 1982, Congress passed the Nuclear Waste Act, which allowed for temporary storage of up to 1900 tons of spent fuel off site, to lessen the threat around nuclear plants. However, this law made no mention of permanent waste disposal — an issue still unresolved as the year 2000 approaches — and didn't mention the persistent problem of what to do with obsolete plants.[12]

Other health problems that surfaced around the time of Three Mile Island concerned several specific categories of people. One was military personnel who had conducted maneuvers after atmospheric bomb tests in Nevada and the Pacific in the 1950s. To mimic use of atomic weapons during a war, soldiers often would wait several minutes after the explosion, then enter areas as close as several hundred yards from the test site, heavily contaminated with fallout. Soldiers did not wear respirator masks during these maneuvers, and in instances such as the August 31, 1957, "Smoky" test, in which large amounts of fallout blew directly toward the troops, soldiers were forced to abandon their mission and retreat in a haze of radioactive dust and smoke.[14] For over 20 years, no attention was paid to any effects on the health of these men from radiation exposure. As the years went on and more of these "atomic veterans" fell ill or died, they began to complain to authorities. In January and February 1978, Congressman Paul Rogers' House subcommittee held hearings on the plight of the former soldiers (including testimony from pioneering nuclear physicist Karl Z. Morgan that there is "no safe level of exposure"),[15] and federal health authorities soon began to collect data. In 1979, the U.S. Centers for Disease Control had located eight soldiers of the 2000 present at the "Smoky" blast who had developed leukemia, significantly greater than the expected number of 3.5. In addition, 111 soldiers had developed cancer, already higher than the 101 expected. The knowledge of what happened to the 200,000 to 500,000 atomic soldiers is still evolving, since the lag between radiation exposure and disease onset may take many years.[14]

Along with soldiers who were in close proximity to above-ground bomb blasts, the study of effects of test fallout on the civilian population blossomed almost simultaneously. Concern among residents of southern Nevada and Utah had existed for many years, but these had been squelched by the iron hand of the AEC. As early as 1965, the Public Health Service's Edward Weiss wrote an internal report for the Commission showing 28 leukemia deaths (vs. 19 expected) in Washington and Iron Counties in southwest Utah.[16] The AEC suppressed the report until it was forced to make it public in 1979 to comply with a Freedom of Information Act request (FOIA was one of the anti-corruption laws enacted just after Watergate). Later that year, the University of Utah's Joseph Lyon published a study showing an even greater gap in leukemia deaths (32 actual vs. 13 expected) from 1959 to 1967. Although the article was published by the prestigious *New England Journal of Medicine*, federal officials lashed out at the findings. Charles Land of the National Cancer Institute wrote in *Science* magazine that the analysis was skewed by a very low leukemia rate in Utah in the 1940s due to a lack of available physicians and misdiagnosis of leukemia as another cause of death.[14] Just as Land's critique was published, more damning evidence appeared in the form of an article by Carl Johnson on cancer *cases*, not deaths, a more inclusive category. Among Mormons in southwestern Utah, the 288 cancer cases Johnson uncovered were 60% *greater* than the expected number of 179. Of particular note were the especially large gaps for radiation-related cancers such as thyroid cancer (20 actual cases vs. 3.1 expected) and leukemia (31 cases vs. 7.0 expected).[17]

Another downwind study in the period concerned those living near a nuclear weapons site. In 1979, Johnson, who was the director of the Jefferson County, Colorado, health department, published data on elevated cancer rates in the county, located just downwind of the Rocky Flats plutonium production complex. While there were 24% more cancers (all types) than would normally be expected (within 13 miles of the plant), even greater excesses were documented by Johnson for testicular, throat, liver, lung, and colon cancer, along with leukemia. Johnson's calculation of 501 excess cancer deaths from exposure to Rocky Flats emissions was a far cry from the estimate of 1 excess cancer death in a 1977 federal study.[18]

Yet another category of humans who received closer scrutiny during this time was nuclear workers. The pioneering analysis of Hanford workers by Mancuso and his colleagues in the mid-1970s sparked similar inquiries. Even the federal government, which had tried its best to suppress Mancuso's work, made patterns of excess disease rates public. Specifically, a 1980 Energy Department study showed workers at the Lawrence Livermore nuclear weapons laboratory in California had developed melanoma, or skin cancer, at five times the rate of the local population.[19] Other internal DOE studies completed in 1984 found elevated cancer cases and respiratory diseases at many of the weapons plants.[20]

With the evidence mounting, potential victims and other concerned citizens took to the courts for compensation and reduction of the atomic threat. The legislative process had proved fruitless, as Congress failed to pass bills proposed by Senators Edward Kennedy and Orrin Hatch compensating atomic fallout victims in 1979, 1981, and 1983.[16] In December 1978, Stewart Udall, an attorney and former Secretary of the Interior under Lyndon Johnson, filed $100 million in claims on behalf of western "downwinders" against the U.S. Department of Energy. After over five years of legal wrangling, a federal judge awarded 10 victims $2.66 million in 1984, the first official punishment dealt to the public–private nuclear establishment.[16] Other legal efforts were not as successful; in 1986, a U.S. Court of Appeals reversed a judge's order granting a new trial to Utah sheep herders who lost many of their flock after a 1953 bomb test, despite the large amount of new evidence indicating radiation caused the sheep to sicken and die.[21]

Another legal effort to stop nuclear proliferation occurred in 1978 when Jeannine Honicker, an anti-nuclear activist from Nashville, filed a suit seeking an injunction to shut down the nuclear fuel cycle in all American nuclear power plants. Honicker's efforts were perhaps inspired because her teenage daughter suffered from leukemia. This ambitious law suit was pursued because of the ongoing reticence of Congress or state legislatures to ban or impose moratoriums on nuclear reactor construction or licensing. The judge listened to testimony from Gofman and Sternglass, who presented evidence for the "supralinear" dose response curve, meaning that even low-level emissions were harmful, but decided to dismiss the case anyway.[22]

Whether or not the growing public flap was the cause, more scientists became bolder about making statements about the harm caused by radiation. In 1979, the third report of the Committee on the Biological Effects of Ionizing Radiation (BEIR III) made the strongest official estimate of radiation-induced cancer to date. The BEIR III panel, convened by the National Research Council, predicted that one half of one percent, or 1 million Americans, would eventually develop cancer from man-made radiation, and suggested 6000 babies would be born with radiation-related deformities each year. The dissention on the committee was strong and vocal. Harold Rossi of Columbia University asserted that "the available data relating to cancer incidence …are either misleading or unreliable" and thus "the BEIR III report will contribute to excessive and potentially detrimental apprehension over radiation hazards."[23] The following year, BEIR III was revised to reflect a lower estimate of radiation-induced cancer. However, the document became the strongest official acknowledgment to date of the health effects of low-level radiation.[24]

Despite the heightened awareness in the late 1970s and early 1980s, the presence of man-made radiation in the environment increased, raising the potential for harm to humans. Production of nuclear weapons rose drastically after Reagan became President, even though the number of tests in Nevada remained relatively steady, at about 20 per year. Although nuclear power plants were no longer being ordered and utilities were rapidly cancelling recent orders, many previously ordered plants began operations, raising the total to new highs. By January 1, 1979, 70 reactors were in operation at 48 sites (some sites had more than one reactor). About 20 million Americans, or 10% of the population, lived within 30 miles of a reactor, a number that would grow further in the next decade as more plants opened.[25]

Did more reactors mean more exposure and more potential harm? The answer to this question is unclear. In the 1970s, reported airborne releases of radionuclides such as iodine-131 grew as more plants opened. The official 1972 U.S. total (from the Nuclear Regulatory Commission) of 20 curies rose to 37 in 1975, 33 in 1977, and 38 in 1979, including 14.2 from the Three Mile Island accident. In 1980, however, the reported figure fell to about 6, and remained below 10 thereafter.[4]

However, data on exposure to workers in nuclear plants contradict this drop. The U.S. General Accounting Office, acting on a request from Senator John Glenn, found that between 1969 and 1980, the average rems absorbed by nuclear workers in a typical reactor rose from 178 to 791.[26] The 1980 figure represented an all-time record, a 35% increase from 1979, even though airborne particulate releases fell dramatically that year.[27] Whether the numbers are all accurate, or whether atomic workers were not protected as well as the public remains a puzzle.

More troubling health and safety data are those on plant reliability. In 1983, the 79 reactors in the U.S. only produced 58% of maximum power, despite the fact that most of them were relatively new (under 15 years old).[28] The 58% figure is typical for the performance of reactors during the last several decades. In addition, the oldest reactors were beginning to show wear and tear; their vessels were turning brittle from years of radiation bombardment, much earlier than expected. This development suggested that radiation leakages were more possible unless plants were closed or repairs were promptly made earlier in their existence.[29]

If the U.S. nuclear industry was slowly sagging through the 1980s, it received a powerful one-two punch, crippling both its military and civilian components, in the middle of the decade. Many believe the industry will never recover completely after these two blows. Oddly, these powerful jolts did not come from within the U.S., but from its long-time nemesis, the Soviet Union.

The first quake occurred in 1985, with the accession to power of Mikhail Gorbachev as chairman of the Communist party. Gorbachev took office in the midst of a renewed Cold War with the U.S. His atomic-era predecessors (Stalin, Khrushchev, Brezhnev, Chernenko, and Andropov) had stuck to a steady course of nuclear arms buildup regardless of the consequences to the nation and the world. Health effects of radiation in the U.S.S.R., a totalitarian police state, were not for public debate. Brave isolated individuals who dared speak out, like Sakharov, were suppressed, exiled, or sometimes killed. The

Soviet arms buildup continued, even though the rest of the nation's economy weakened. When commercial nuclear power became available, the Communist hierarchy eagerly bought into it with little regard for safety and health.

Gorbachev, however, represented a change from the past. Unlike previous Soviet leaders, he was an educated, worldly man, who understood the West and international politics. Not having lived under the Czarist regime and the aftermath of the 1917 revolution, Gorbachev offered talents as a skilled bureaucrat rather than a fiery demagogue. Although he was strongly committed to civilian nuclear power, vowing to double the Soviet atomic energy capacity, he saw the heavy toll the arms race was taking on Soviet society, and took immediate action to reverse previous policies.

In September 1985, Gorbachev met with Reagan for the first time, just six months after he assumed his position. The first private exchange between an old American Cold Warrior and the heir to a hard-line police state was a hint that things were about to change. In front of only the translators, Reagan spoke to Gorbachev:

> Here you and I are...probably the only two men in the world who could bring about World War III. But by the same token, we may be the only two men in the world who could perhaps bring about peace in the world. We can keep them waiting out there as long as we want. We're in charge. It's up to you and me.[30]

Although the Geneva meeting made no great breakthroughs in arms limitations, it marked the beginning of the path to a new understanding; the first treaty, to eliminate 2600 short- and intermediate-range missiles, was signed in December 1987.

If the "new" Soviet Union was a clap of thunder that changed the nuclear course in America, then the bolt of lightning struck on April 26, 1986. In the Ukrainian town of Chernobyl, situated on the border of Belarus 60 miles north of Kiev, stood a complex of nuclear reactors generating electricity for the region. The surrounding area had many farms, which is why the Ukraine long held the reputation as the Soviet Union's "breadbasket." About 110,000 people lived within 30 kilometers (18.5 miles) of the plant.[31]

Just after midnight on April 26, a test was in progress at reactor #4, which had only opened in December 1983. The test was an attempt by plant operators to see if the turbines could continue to produce electricity even if the reactor was no longer powering them; however, operators dropped the megawattage to such a low level, that the flow of steam to generators stopped and the reactor became impossible to control. The evaporated fuel built up into an enormous mass of steam, making a disaster imminent. At 1:24 a.m., several explosions took place, completely blowing the massive, 2000-ton concrete roof off of reactor #4, and allowing at least 90 million curies of radionuclides to escape into the environment from the white-hot, now exposed core.[32] Eyewitnesses, such as G. N. Petrov, who was driving his car toward the nearby town of Pripyat, described the horrifying sight:

> I approached Pripyat around 2:30 a.m. from the north-
> west, from the direction of Shipelichi. From Yanov sta-
> tion I could already see the fire above No. 4 unit. The
> ventilation stack, with its horizontal red stripes, was
> clearly lit up by the flames. I remember how the flames
> were higher than the shaft, so that they must have been
> nearly 600 feet in the air...by the light of the fire, I
> could see that the building was half destroyed.[31]

The health effects caused by Chernobyl were staggering, leading some to predict that the eventual number of victims would exceed those of Hiroshima and Nagasaki. The first to succumb were the brave Soviet workers who struggled to put out the fire (accomplished May 6) and bury the dam-aged reactor in a sarcophagus of concrete; 31 of these workers died within days.[33] By 1991, the Soviet government admitted to a death toll of 254 from acute radiation sickness, a figure that many in the former Soviet Union ridicule as being extremely conservative.[34] By 1996, 30,000 of the 400,000 working to control the damage had fallen ill, 5000 of whom became unable to work.[32] The number of local victims went far beyond those suffering from instant radiation sickness from massive doses; thousands living near the plant (especially downwind) experienced symptoms, disease, and death in far greater numbers than normal. Although records have not been kept with great precision by Soviet health officials, and it has been only 10 years since the accident, the toll is astounding. Figures obtained by Kiev municipal official Leonid V. Skripka from the KGB (Soviet secret police) show a massive increase in a variety of diseases among 30,000 local residents from 1987 to 1991:

- Overall death rate   400% increase
- Total cancer rate   300% increase
- Breast cancer rate   26% increase
- Brain cancer rate   300% increase
- Overall disease rate   500% increase
- Adult pneumonia rate   220% increase
- Child pneumonia rate   260% increase

Other diseases with reported rises include pulmonary ailments, colds, influ-enza, complicated pregnancies, maternal deaths, underweight births, birth defects, allergies, immune suppression from low natural killer cell numbers, impotence, low sperm counts, and neuropsychiatric disorders.[34] Greenpeace estimates 32,000 Ukrainians have died from 1986 to 1996 as a result of Chernobyl.[32] Naturally, Chernobyl fallout spread to neighboring European countries as it surged beyond the Soviet borders. The Otto Hug Radiation Institute in Munich measured the amount of strontium-90 in baby teeth of West German children. Baby teeth from persons born in 1987, when atmospheric strontium was still entering the diet from the atmosphere via precipitation, showed 1.04 picocuries of strontium-90 per gram of calcium, nearly 10 times that of children born between 1983 and 1985 (0.11), prior to the Chernobyl accident.[35]

The fallout from Chernobyl entered the U.S. atmosphere about May 5. Because it had taken over a week to travel, many of the short-lived radionuclides had decayed, and because of the great distance from Chernobyl, the concentrated mass of radioactivity that horribly gripped the area near the reactor had thinned out. However, Chernobyl radioactivity made its way into the American food chain in amounts much larger than routine levels, but American health officials believed these amounts were too minuscule to issue any warnings. The average concentration of iodine-131 (with a half-life of 8 days) in U.S. milk was about 13 picocuries per liter in May 1986, followed by about 9 in June, before returning to customary levels of about 2 or 3, according to the Environmental Protection Agency's monthly report *Environmental Radiation Data*. Cesium-137 averaged about 10 picocuries per liter in the last eight months of 1986; even the longer half-life (30 years) of this chemical made the return to customary levels of 2 or 3 last into 1989.[36]

All of the above measurements were taken by the EPA, a federal agency. However, state health departments also made readings, sometimes finding amounts of radioactivity well above the EPA's. The Washington state Department of Health and Social Services measured an iodine-131 level of 560 *picocuries per liter* of milk in Redland on May 5.[37] Populated areas also experienced discrepancies. In New York City, the peak levels of iodine-131 and cesium-137 (82 and 80) found by local officials were far greater than those reported by the EPA (32 and 21).[38]

It wasn't long before health effects of this added radiation to the American diet began to surface. In 1989, Jay Gould and Ernest Sternglass published a paper showing that during the four months after Chernobyl, the U.S. experienced about 15,000 more deaths than in the same period of 1985, and 5,000 more deaths from September to December 1986. Gould and Sternglass showed the chances of this 15,000-person excess of deaths in four months had never happened since the great Spanish flu epidemic in 1918.[39] They reasoned that the added insult of low levels of radiation hastened the deaths of sick and otherwise susceptible persons. Although the two scientists were criticized by some for concluding that these deaths were connected with Chernobyl fallout, critics presented no alternative explanation and could not argue with the statistical power of the discovery.

Information produced by the National Center for Health Statistics' annual volume *Vital Statistics of the United States* appears to uphold the Gould/Sternglass contention that some unusual event caused an abrupt jump in U.S. mortality in the summer of 1986 (Table 4.3). In addition to deaths

Table 4.3 Percent Change in Death Rates from Previous
Year in the U.S., 1986

	Jan–Apr	May–Aug	Sep–Dec
All Causes			
Total population	–0.2	+2.3	+0.8
Under 1 year	–2.3	+0.1	–6.2
Congenital Anomalies			
Total population	–4.4	+4.0	–2.8
Under 1 year	–7.1	+1.5	–5.6

from all causes, deaths from congenital anomalies (birth defects) and deaths to infants under one year showed similar patterns.

After 1990, the first documentation of disease-specific health effects of Chernobyl appeared in several peer reviewed medical journals. The first showed an enormous rise, as much as 100 times, in thyroid cancer in the former Soviet states of Belarus and the Ukraine.[40-42] But other articles began to show that changes that would reflect an adverse reaction to radioactivity from Chernobyl had occurred in the U.S. as well.

Thyroid Cancer. In 1995 and 1996, reports of a corresponding rise in thyroid cancer in several American states also appeared, not just among children but for the entire population. Between the late 1980s and early 1990s, the age-adjusted thyroid cancer rate rose in Connecticut, Iowa, and Utah by 19.4%, far exceeding the 6% increase recorded in the 1980s. This change was much smaller than those near Chernobyl, but fit the same pattern of thyroid cancer cases beginning an unexpected increase four years after the explosion.[43,44] Moreover, it lent credence to the supralinear dose response curve, where low levels of protracted radiation exposure could be harmful.

Newborn Hypothyroidism. Another potentially radiation-related disorder due to ingestion of iodine, newborn hypothyroidism (underactive thyroid gland), also increased after Chernobyl. The U.S. rate rose 8.3% between 1984–85 and 1986–87, after virtually no change in the late 1970s and early 1980s. Furthermore, the area with the greatest concentration of thyroid-seeking iodine (the northwest U.S.) experienced a 23.3% rise, while the region with the least fallout (the southeast U.S.) saw its rate decline by 1.0%.[45]

Infant Leukemia. Emulating similar patterns in Greece and the former West Germany, leukemia cases diagnosed before the first birthday in American children born in 1986 and 1987 was 30% higher than for children born in the remainder of the decade.[46] The 30% excess was smaller than those found in Greece (160%)[47] and West Germany (48%),[48] but these nations were hit much harder than the U.S. by Chernobyl fallout, and the U.S. study was based on many more cases than the other two nations. Findings of excess thyroid cancer, newborn hypothyroidism, and infant leukemia are a start, but it will probably take well into the 21st century to fully understand the many effects of Chernobyl, in the U.S. and around the world.

After Chernobyl, almost all developments represented bad news for the American nuclear enterprise. Perhaps atomic weapons production suffered the fastest and the most. Partly as a result of Chernobyl, the U.S.S.R. began to crumble in the late 1980s, and officially disbanded in 1990 into a league of autonomous states. The Communist party gave way to other political parties in the new states, and freedoms — speech, press, open elections, etc. — were introduced into the society. The Soviets relinquished their military and economic grip on Eastern European nations, and these countries sought to introduce private enterprise into their economies, leaning particularly hard on American capitalists. To free up capital and to induce American cooperation, Russia and other former Soviet republics signed several disarmament treaties during these years, and nuclear weapons were dismantled rapidly. The agreements envisioned the U.S. and Russia with about 3000 atomic weapons each by 2003, down from a high of over 30,000 apiece. The Cold War was over.

Because the end of the Cold War meant the end of the arms race, American nuclear weapons plants were suddenly in trouble, having outlived their usefulness. The government became more open in publicizing the health hazards these plants presented. Hanford, one of the two oldest plants and the one with perhaps the "dirtiest" past, was the first to come under public scrutiny. In February 1986, the Energy Department released 19,000 pages of documents, carrying with them some stunning news. In the late 1940s, hundreds of thousands of curies of iodine-131 had been released, including over 400,000 curies of iodine-131 to manufacture plutonium for the Alamogordo and Nagasaki bombs in 1945. A deliberate and secret release on December 3, 1949, of 5000 curies of iodine-131 known as the Green Run, a test of conditions in the aftermath of a nuclear attack, was another shocking revelation. Memoranda acknowledging government knowledge of the danger of these and other releases were included in the documents.[49] The bad news about Hanford's past did not subside, and one by one, all its reactors were closed, permanently, until weapons production had ceased by the end of the decade.

Studies commenced on health effects to both workers and the general population. In 1988, DOE authorized the Hanford Dose Reconstruction Project to calculate the types and volume of radiation received by the people in eastern Washington state, an effort that had yet to produce results by early 1998.[49] In June 1990, after years of resistance, DOE finally agreed to release the previously secret health records of 50,000 workers at Hanford, plus several hundred thousand more from other weapons production plants.[50] Data were handed over to Alice Stewart, by now in her 80s, for independent analysis. Stewart's initial work showed that 3% of cancer deaths to Hanford workers between 1944 and 1986 were due to radioactive emissions. In particular, exposures to workers over age 50 were harmful at lower doses.[51] Perhaps most distressing, the threat at Hanford was far from over. Billions of gallons of chemical waste had been poured into the ground at Hanford over the years and could not be contained.[49] Furthermore, 67 of 177 underground tanks containing waste are believed to be leaking, and contaminants have reached a level 230 feet below the tanks.[52] In particular, the plutonium in the Hanford soil, at concentrations up to 30 times more than the area just 20 miles to the southeast, represented a serious health hazard, since the isotopes like plutonium-239 have very long decay processes (half life of 24,000 years).[53] Future seepage of radioactive waste through the groundwater and into the Columbia River will not take place in the near future, but in the next century, this development will threaten hundreds of thousands of lives in the Pacific northwest if the present course is not reversed.[54]

The Savannah River plant near Aiken, South Carolina, was the next to melt in the bright public lights. In September 1988, Senator John Glenn, a long-time skeptic of the Energy Department's nuclear weapons policies, held hearings that disclosed for the first time that two nuclear rod meltdowns had occurred at Savannah River in late 1970. The cult of secrecy that prevailed in the weapons program kept the accidents from the public for 18 years, despite the threat to the public. Similar to Hanford, nuclear weapons production at Savannah River was soon suspended. The plant was also saddled with the problem of what to do with contaminating nuclear waste. Savannah River and the nearby Barnwell site are regional waste repositories, storing about half of the waste from nuclear weapons plants in the U.S.[55]

Other weapons plants suffered from the new openness as well. At Oak Ridge, Tennessee, public pressure forced the state health department to establish a Health Assessment Steering Panel. The group's first report in 1993 was a compendium of radioactive emissions and other contamination from the plant's operations; and thereafter, the panel embarked on an estimate of actual dosages to the local population.[56] Helping to bring change at Oak Ridge was a 1991 study by Stephen Wing showing a leukemia death rate in plant workers 63% above expected (37 deaths vs. 22 expected). For all cancers combined, a 47% excess (22 deaths vs. 15 expected) occurred for those receiving over 4 lifetime rems of exposure from the plant.[57] Four rems is equal to the amount of radioactivity each American receives from background radiation exposure from rocks, soil, etc. over a 40-year period.

Workers at the Feed Materials Production Center in Fernald, Ohio, were found to have excessively high death rates for a variety of cancers. The study was an independent 1994 effort by Professor Peter Gartside of the University of Cincinnati for a lawsuit against the uranium processing complex.[58]

Perhaps the final strike against the nuclear weapons program occurred in December 1993, when DOE admitted 204 earlier atomic bomb tests, about 20% of all tests, had been kept secret from the public.[59] By the early 1990s, the American nuclear weapons program was in a sorry state. All weapons production operations had been suspended. Weapons testing had ceased (the final Nevada test occurred on September 23, 1992).[59] Some weapons plants now concerned themselves with research (at places like Oak Ridge) and weapons disassembly (at Pantex in Amarillo, Texas), but the number-one purpose of all these sites remains the control of contamination from earlier radioactive emissions and nuclear waste.

Table 4.4   Death Rate, All Cancer, % Above/Below U.S. Rate
in Counties Near Nuclear Weapons Plants

Site, State (Counties)	Opened	% Above/Below U.S.		% Change
		Early '50s	Early '80s	
Hanford, WA (3)	1943	−22	−6	+16
Oak Ridge, TN (2)	1943	−8	−9	−1
Mound, OH (3)	1947	+1	+11	+10
Idaho Nat'l, ID (3)	1949	−22	−32	−10
Ballard, KY (2)	1950	−15	+8	+23
Savannah R., SC/GA (3)	1950	−12	−1	+11
Fernald, OH (2)	1951	+12	+19	+7
Portsmouth, OH (1)	1952	−12	+18	+30
Rocky Flats, CO (2)	1953	−11	−5	+6
Nuc. Fuel Svcs., NY (1)	1966	−3	+8	+11

Atomic power plants also had a tough time after Chernobyl. The public outcry was moving local health departments to take a harder look at the health effects of radioactive emissions. In May 1987, the Massachusetts Department of Public Health found a 50% higher rate of various blood cell cancers, including leukemia, near the troubled Pilgrim reactor.[60] The same year, prodded by a group of mothers, officials in Lincoln County, Maine, found significantly higher rates of breast, genital, and cervical cancer near the Maine Yankee plant.[61]

Probably the biggest revelation about effects from nuclear power plants in the late 1980s came in 1988, when Senator Kennedy directed the National Cancer Institute to undertake a comprehensive study of cancer mortality in the areas near nuclear reactors. Two years later, NCI produced a massive report which stated, "the survey has produced no evidence that an excess occurrence of cancer has resulted from living near nuclear facilities."[62] However, the details of the thick, three-volume report often did not support this conclusion. The NCI failed to note, for example, that cancer mortality rates in the early 1980s (the latest period covered by the study) were at their highest level in many locations, especially those near nuclear weapons plants that opened 30 to 40 years earlier. Because cancer mortality often has a long period after exposure to radiation, and because the weapons plants had the highest levels of emissions, especially in the 1940s and 1950s, this finding is important, but was ignored by the NCI. Table 4.4 shows the rise in all nuclear weapons sites covered in the analysis.

For 8 of the 10 sites, comprising over 6000 cancer deaths per year by the early 1980s, the local mortality rate increased faster over the 30-year period compared to the national change; an approximate rise of 10% means that in these 22 counties alone, *600 cancer deaths a year* are excessive, and linked to the weapons plants. The NCI failed to take note of these findings. It did say, however, that analyzing cancer *deaths* is not as meaningful as cancer *cases.* Many types of cancer can be cured or controlled, so mortality statistics may hide the full effects of radiation exposure. To its credit, the NCI added information on cancer cases for plants in Connecticut and Iowa, which have reliable, established cancer registries. It acknowledged that leukemia cases occurring in children jumped after power plants began operations. However, the report again failed to assert that unusual increases took place for many other cancers besides leukemia. Significant increases are clear for both Connecticut plants, which have been in service the longest, but not for that in Iowa (only operating 10 years by 1984, the latest year covered in the study).

The post-Chernobyl clamor over safety and health near nuclear plants reached the BEIR committee. In the group's 5th report late in 1990, it recommended that the maximum annual exposure for nuclear workers be dropped from 5 to 2 rems. It also asserted that up to 800 cancers may develop per million people from exposure to 1 rem of radioactivity. However, the committee stopped short of buying into the supralinear curve concept and would not admit that low doses of routinely emitted radionuclides were more harmful than previously thought.[63]

Compounding the lowered public perception about radiation was the befuddled attempt to house high-level atomic waste in a permanent location. The topic had been discussed for decades, but no permanent solution had ever been agreed on, and waste piled up in temporary locations around the county. By the early 1990s, these locations were filling up, and people living near centralized sites, such as the one in South Carolina, increasingly began to resent what they perceived as a growing menace. The Energy Department proposed a permanent waste repository in the volcanic rock deep below the Yucca Flats area in Nevada. However, plans have stalled because a number of scientists aren't convinced that the site will be stable enough to protect the material as it decays over thousands of years.[64]

The future course of U.S. nuclear weapons manufacturing, power production, or waste management are all uncertain. Much of it depends on external factors. For example, weapons production will be governed by the stability of the post-Cold War world, especially in American relations with the former Communist states, and with a variety of third-world countries. Nuclear power production depends on the availability and feasibility of alternative power sources, and on the profitability of nuclear power itself. Key to each of these areas is the potential threat to safety and health, both in terms of what scientists say it is and what the public believes it is. No one can tell for sure what the long-term future will hold. But the mounting body of evidence that radiation is harmful even in low doses; the revelations of Cold War deceptions on nuclear safety (the latest being the 1994 articles by the Albuquerque *Tribune* on experiments with radioactive substances on 16,000 Americans)[65]; the long memories of Three Mile Island and Chernobyl; and the continuing problem of waste disposal and contamination make the possibility of future growth doubtful, at best.

In 1996 and 1997, just before publication of this book, the nuclear industry continued to be battered by new findings of health and safety violations. The Brookhaven National Laboratory on Long Island, which has operated a research reactor for the Energy Department since 1950, ran into trouble in

1997. The announcement that radioactive products, especially tritium, had leaked out of the plant for years into the area's groundwater, along with airborne releases, was the lightning rod. The coalition of universities that ran Brookhaven had known about the contamination since 1985, and had basically ignored the problem. Jay Gould quickly pointed out that Suffolk County (the home county of Brookhaven) had the highest breast cancer death rates in the country, and the five towns closest to the reactor had the highest rates in the county. Brookhaven and its parent Energy Department tried to fight back: it offered free connections to the public water supply to residents of a large area south of the plant, because these residents felt their wells were contaminated.[66] It formed a public advisory committee, and it terminated the contract of the universities that had operated the plant, a group which included prestigious Ivy League schools, before replacing them with a local university.[67] Still, the probe of health problems caused by Brookhaven continues. In 1997, Gould's Radiation and Public Health Project began to collect baby teeth from local children, to measure the concentration of strontium-90 in the teeth.

Another recent action against nuclear reactors based on health concerns occurred in March 1996. A cover story appeared in *Time* magazine, detailing a series of safety violations in the four nuclear reactors in Connecticut that had gone unaddressed by plant operators, Northeast Utilities. Among the more serious of these charges was the underwater storage of highly radioactive spent fuel rods in numbers greater than permitted by federal regulations, which attempt to minimize the potential damage if the "hot" fuel rods were ever taken out of cooling water. Within weeks, panicky officials from the Nuclear Regulatory Commission closed the single reactor at the Haddam Neck plant in East Haddam along with the three reactors at the Millstone plant near New London. Northeast Utilities elected to permanently close the 29-year-old Haddam Neck reactor in December 1996, and by early 1998, the three Millstone reactors were still closed, pending further NRC review. Federal regulators also assessed a whopping $2.1 million fine on Northeast Utilities for violations at Millstone, by far the largest penalty to date.[68]

The closing of Haddam Neck may have signaled the demise for the many reactors that are now nearing 30 years in operation. In April 1997, General Public Utilities Corporation, the operator of the Oyster Creek reactor in central New Jersey, announced it was exploring options to sell or close the 28-year-old facility by the year 2000.[69] The utility's stated reason for considering these new options was the high cost of running the plant. But it neglected to mention the plant's long history of safety and health problems. In the 1970s, Oyster Creek emitted more airborne iodine and other short-lived radioactive products than any U.S. plant except the Dresden reactor in northeastern Illinois.[4] In 1997, a local group of citizens publicized the high rate of childhood cancer in nearby (and downwind) Tom's River, and enlisted the state health department to provide technical assistance. The NCI study also shows that cancer rates in Oyster Creek's home county of Ocean County had risen 7% from the late 1960s (before the plant opened) to the early 1980s (the latest data used by the NCI). This amounts to an excess of 82 cancer deaths each year in a modestly sized county (350,000 inhabitants in the early 1980s).[62]

New revelations of radiation's health effects also included updates on bomb test fallout. On August 1, 1997, the National Cancer Institute released a study estimating radioactive iodine-131 doses received during the 1950s for residents of each of the 3000-plus counties in the U.S. The research began in 1982, and was largely forgotten as the work dragged on. Even after the study was completed in 1994, the NCI inexplicably held onto the information for three more years. When released, however, the study reverberated throughout the scientific, political, and media communities. The NCI estimated that bomb blasts caused between 10,000 and 75,000 thyroid cancer cases, three fourths of whom were children under five in 1952 to 1957.[70] The NCI study was its first admission that bomb testing had caused harm in the form of a specific disease, and was its first to point out that fetuses, infants, and children were most affected, due to *low-dose* fallout across the U.S.

The nuclear industry may be down in the late 1990s, but certainly it is not to be counted out. It has continued its ability to avoid liability in court cases. In June 1996, federal district judge Sylvia Rambo dismissed the suits of over 2000 Pennsylvania residents suffering from immune-related diseases, which claimed that they had been harmed by the Three Mile Island accident. All the plaintiffs had cancers and leukemias, but failed to sway the judge.[71] By the end of 1997, domestic production of new reactors in the U.S. was a dead issue; no reactors had been ordered since 1978 (and no order since 1973 resulted in a reactor that was actually built), and no prospects were in sight. However, the industry was bolstered in November 1997 by the news that the Clinton administration had agreed to sell nuclear power reactors to China.

As part of the "dose-response" analysis that scientists use in radiation studies, these past three chapters have addressed the "dose" aspect, a rundown of sources and amounts of man-made radioactivity added to the environment since the 1940s. The next five chapters will constitute the "response" portion, a look at any changes in human health that may be due to radiation exposure. If changes are consistently adverse ones, a statistical link is made, and the words "cause and effect" can be used. We will take a look at two specific populations — those born between the mid-1940s and mid-1960s, and those born since the early 1980s — and present data that shed light on how these populations may have been affected by radioactive chemicals produced in the past 50 years.

References

1. Ford, D., *Three Mile Island: Thirty Minutes to Meltdown*, Penguin Books, New York, 1981, 16-17, 217-247, 258.
2. *The New York Times*, May 7, 1979, 1.
3. Sternglass, E., *Secret Fallout: Low-Level Radiation from Hiroshima to Three Mile Island*, McGraw-Hill, New York, 1981, 204-205, 262, 265-266.
4. Tichler, J., et al., *Radioactive Materials Released from Nuclear Power Plants: Annual Report 1993*, Brookhaven National Laboratory, Upton, NY, 10-13.

5. *The New York Times*, April 5, 1979, II 14.
6. MacLeod, G., A role for public health in the nuclear age, *American Journal of Public Health*, March, 1982, 237-239.
7. *The New York Times*, February 21, 1980, 16.
8. *The New York Times*, September 6, 1985, I 10.
9. Procter, R., *Cancer Wars: How Politics Shapes What We Know & Don't Know About Cancer*, Basic Books, New York, 1995, 56.
10. Epstein, S., *The Politics of Cancer*, Anchor Books, Garden City, NY, 1979, 20.
11. Schieffer, B. and Gates, G. P., *The Acting President*, E. P. Dutton, New York, 1989, 327.
12. Hertsgaard, M., *Nuclear Inc.: The Men and Money Behind Nuclear Energy*, Pantheon Books, New York, 1983, 160, 210.
13. Clement, F., *The Nuclear Regulatory Commission*, Chelsea House, New York, 1989, 45.
14. Rosenberg, H., *Atomic Soldiers: American Victims of Nuclear Experiments*, Beacon Press, Boston, 1980, 123.
15. *The New York Times*, January 25, 1978, A14.
16. Ball, H., *Justice Downwind: America's Atomic Testing Program in the 1950s*, Oxford University Press, New York, 1986, 89, 114, 116, 133, 154, 171.
17. Johnson, C., Cancer incidence in an area of radioactive fallout downwind from the Nevada test site, *Journal of the American Medical Association*, January 13, 1984, 230-6.
18. *The New York Times*, April 10, 1979, A16.
19. *The New York Times*, April 27, 1980, 52.
20. High cancer rates found in nuclear plants. *New Scientist*, October 1, 1984.
21. Udall, S., *The Myths of August*, Pantheon Books, New York, 1994, 216.
22. *Shut Down: Nuclear Power on Trial: Experts Testify in Court*, The Book Publishing Company, Summertown, TN, 1979.
23. *The New York Times*, May 3, 1979, IV 6.
24. National Research Council, *Health Effects of Exposure to Low Levels of Ionizing Radiation (BEIR III)*, National Academy Press, Washington, DC, 1980.
25. *The New York Times*, May 6, 1979, IV 6.
26. *The New York Times*, October 11, 1982, 16.
27. *The New York Times*, September 3, 1981, B12.
28. *The New York Times*, September 16, 1984, IV 6.
29. *The New York Times*, September 27, 1981, 1.
30. Sheehy, G., *The Man Who Changed the World*, Harper Collins, New York, 1990, 193.
31. Medvedev, G., *The Truth About Chernobyl*, Harper Collins, 1991, 31, 88.
32. Shcherbak, Y., Ten years after the Chernobyl era, *Scientific American*, April 1996, 45-46.
33. Gale, R. and Hauser, T., *Final Warning: The Legacy of Chernobyl*, Warner Books, New York, 1988, 175.
34. Cheney, G., *Chernobyl: The Ongoing Story of the World's Deadliest Nuclear Disaster*, New Discovery Books, New York, 1993, 56, 62, 67.
35. Scholz, R., *Ten Years After Chernobyl: The Rise of Strontium-90 in Baby Teeth*, Radiation and Public Health Project, New York, 1997, 9.
36. Environmental Protection Agency, *Environmental Radiation Data*, Montgomery, AL, 1986, Reports 45-48.

37. Washington Department of Health and Social Services, DHHS activities relating to the Chernobyl nuclear accident, Olympia, WA, 1986.
38. Department of Energy, Concentrations of I-131, Cs-134, and Cs-137 in milk in the NY metropolitan area following the Chernobyl reactor accident, New York, Report #EML 460, 308-326.
39. Gould, J. and Sternglass, E., Low-level radiation and mortality, *CHEMTECH*, January 1989, 18-21.
40. Kazakov, V. S., Demidchik, E. P., and Astakhova, L. N., Thyroid cancer after Chernobyl, *Nature*, September 3, 1992, 21.
41. Likhtarev, I. A., Sobolev, B. G., Kairo, I. A., et al., Thyroid cancer in the Ukraine, *Nature*, June 1, 1995, 365.
42. Stsjakhko, V. A., Tysb, A. F., Tronko, N. D., et al., Childhood thyroid cancer since accident at Chernobyl, *British Medical Journal*, March 25, 1995, 801.
43. Reid, W. and Mangano, J., Thyroid cancer in the United States since the accident at Chernobyl, *British Medical Journal*, August 19, 1995, 511.
44. Mangano, J., A post-Chernobyl rise in thyroid cancer in Connecticut, U.S.A., *European Journal of Cancer Prevention*, March 1996, 75-81.
45. Mangano, J., Chernobyl and hypothyroidism, *Lancet*, May 25, 1996, 1482-3.
46. Mangano, J., Childhood leukaemia in U.S. may have risen due to fallout from Chernobyl, *British Medical Journal*, April 19, 1997, 1200.
47. Petridou, E., et al., Infant leukaemia after *in utero* exposure to radiation from Chernobyl, *Nature*, July 25, 1996, 352-3.
48. Michaelis, J., et al., Infant leukaemia after the Chernobyl accident, *Nature*, May 15, 1997, 246.
49. D'Antonio, M., *Atomic Harvest: Hanford and the Lethal Toll of America's Nuclear Arsenal*, Crown Publishers Inc., New York, 1993, 1, 119-125, 261.
50. *The New York Times*, May 18, 1990, A17.
51. Kneale, G. W. and Stewart, A. M., Reanalysis of Hanford data: 1944–1986 deaths, *American Journal of Industrial Medicine*, 1993, 371-389.
52. *The New York Times*, November 26, 1997, A25.
53. *The New York Times*, April 12, 1985, 85.
54. *The New York Times*, October 11, 1997, A7.
55. Gould, J. and Goldman, B., *Deadly Deceit: Low-Level Radiation, High-Level Cover-Up*, Four Walls Eight Windows, New York, 1990, 40.
56. Oak Ridge Health Assessment Steering Panel, Vols. I-V, *Oak Ridge Health Studies Phase I Overview*, Nashville, TN, 1993.
57. Wing, S., Mortality among workers at Oak Ridge National Laboratory, *Journal of the American Medical Association*, March 20, 1991, 1397-1402.
58. *The New York Times*, April 13, 1994, A17.
59. Norris, R. and Cochran, T., *United States Nuclear Tests, July 1945 to 31 December 1992*, Nuclear Resources Defense Council, Washington, DC, 1994, 1, 53.
60. *The New York Times*, May 21, 1987, 22.
61. *The New York Times*, August 16, 1987, 50.
62. Jablon, S., et al., *Cancer in Populations Living Near Nuclear Facilities*, National Cancer Institute, Washington, DC, July 1990.
63. National Research Council, (BEIR V), National Academy Press, Washington, DC, 1990.
64. *The New York Times*, March 5, 1995, A1.

65. *The New York Times*, August 20, 1995, 27.
66. *The New York Times*, March 22, 1997, 28.
67. *The New York Times*, November 26, 1997, B5.
68. *The New York Times*, December 11, 1997, B6.
69. *The New York Times*, April 11, 1997, B1.
70. *The New York Times*, August 2, 1997, 6.
71. *The New York Times*, June 8, 1996, 11.

part two

Health effects and the baby boomers

chapter five

Health effects and the baby boomers — infancy

The American "Baby Boomer" generation, defined as those born from about the mid-1940s to the mid-1960s, received its name due to the high birth rate during those years. These 75 million Americans grew up in perhaps the most prosperous conditions in world history, maybe even more prosperous than succeeding generations. World War II had ended, ushering in a period of sustained economic growth never seen before. This expansion helped the wealthy few, naturally, but also gave an unprecedented boost to the middle class. Income of the average worker grew steadily. Unemployment was consistently low. Enrollment in colleges swelled, fueled by the federally sponsored GI bill of rights. Car sales to middle-class Americans exploded. The greater number of cars meant a move from apartments in crowded cities to houses in more open suburbs. New developments, such as Levittown, Long Island, in 1947, marked the first of many suburban sprawling housing complexes. Middle-class people didn't rent these places, they began buying houses for the first time, helped greatly by housing loans backed by the Veterans Administration. When television was introduced in the late 1940s, many Americans jumped at the chance to have one of these exciting new inventions. By the 1950s, a large portion of the swelling ranks of middle-class Americans owned a car, a house, and a television, along with other new electronic gadgets for the house. With more money available, people began to expand their leisure activities. Time off from work no longer meant staying home for a week, but instead was a chance to jump in the car or hop a plane and go touring. The federal government helped out again by developing a system of interstate highways in the 1950s. Private chains of hotels, tourist attractions such as Disneyland, and golf resorts sprang up all over the country.

The generation growing up during these times had far more opportunities as youths than their parents or grandparents. They lived in better housing; were educated for a longer period in better schools; wore better clothing; and ate a more nourishing diet. These key components of the living standard are especially helpful to the young, who are developing physically and emotionally. The early years are crucial in determining a person's future well-being, and the Baby Boomers had a great jump from the beginning.

Another aspect of the booming post-war years helping the young to thrive was the medical care system. Good economics meant good medical care. The Hill–Burton Act of 1946 gave federal assistance to hospitals for new construction and expansion; between 1946 and 1970, 334,000 hospital beds and 94,000 nursing home beds were added to the system. Medical schools grew from 79 to 126 between 1950 and 1980, and many of the existing schools expanded their enrollment, unleashing a large new contingent of well-trained doctors to help Americans stay healthy. After 1960, spending on medical research (much of which came from Washington) skyrocketed. All this expansion offered great benefits for the nation's young. More and more working Americans could afford private doctors for their families, largely because of health benefits offered by employers. The poor and their children still had to scramble for access to care, and were often without insurance; but they did benefit from greater charity care mandated by Hill–Burton, and from government programs such as Kerr–Mills. Access to care for poor children was upgraded significantly in 1966, with the passage of the Medicaid law.

Children were the immediate beneficiaries of some of these new technologies. The availability of new antibiotics and other "wonder drugs" provided very effective means to defeat dreaded diseases such as tuberculosis, influenza, and pneumonia. Development of new vaccines gave Baby Boomers a respite from earlier scourges such as polio, measles, and mumps.

With this happy scenario in place, one would expect the health status of young Americans would improve. However, the introduction of nuclear fission products into the environment during the first 20 years after World War II helped negate improvement, and even worsened health status for a number of conditions that depend on the strength of the immune response. The following chapters describe some of the basic health status measures for Boomers. National data are presented, along with information for populations living near nuclear plants or in places heavily exposed from atmospheric atomic bomb testing, to illustrate the link between radiation exposure and harmful health effects.

Unless otherwise noted, all data used in Chapters 5 through 9 are taken from the National Center for Health Statistics' *Vital Statistics of the United States*, published each year since 1937.

Infant mortality

National patterns

Possibly the most basic measure of a society's health is infant mortality, defined as the proportion of babies born alive that die before their first birthday. Local health departments had documented a decline in infant death

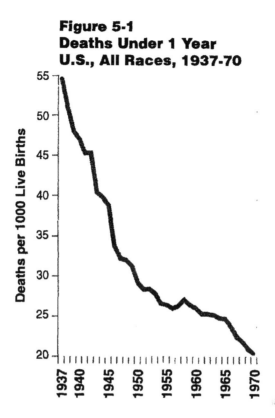

Figure 5-1
Deaths Under 1 Year
U.S., All Races, 1937-70

rates starting in the early 20th century. New York City's rate fell from 14% to 10% between 1898 and 1915. After 1915, when the federal National Center for Health Statistics began collecting infant mortality data from state health departments, the rate continued its steady reduction across the country. The earliest official national tabulation was 10% of all live births in 1915; the rate dropped virtually every year thereafter, even in the years of the Great Depression and World War II. By the mid-1930s, all states were included in federal reports, allowing a true national infant mortality rate to be computed (Figure 5.1). Between 1937 and 1951, rates were cut nearly 50%, from 54.5 to 28.4 per 1000 live births. But suddenly, in 1951, amid continuing economic prosperity and widening access to health care, the decline abruptly stopped. Between 1951 and 1965, the rate only moved from 28.4 to 24.7 per 1000 live births (a 13% drop), the poorest performance in this century. After 1965, the rate continued falling sharply, to a current level of below 10 per 1000 births.

The lack of progress from 1951 to 1965 included both whites (17% decrease) and blacks (6% decrease). In 1952, 1957, 1958, 1962, and 1963 there was no decline, or even an increase. Many of the deaths were to babies less than one month old; this "neonatal" death rate fell just 15% for whites and 4% for blacks between 1951 and 1965, after years of steady declines. The unexpectedly small drop in infant death rates during these years forms the basis for Sternglass' estimate of 375,000 excess infant deaths due to bomb testing.

Radiation-related damage will increase the risk of a baby not surviving its first year of life. Infant deaths are often attributed to prematurity, under-development of crucial organs such as the lungs, congenital malformations, and inability to fight off bacterial and viral infections. Various radioactive chemicals found in bomb fallout can impair physical development, immune system strength, and genetic makeup. Direct exposures to the infant, or those passed on to the fetus by either parent can put a baby at greater risk of not surviving its first year.

The national infant mortality rate is a very broad measure, and it is difficult to attribute its pattern to only one cause. However, the fact that President Truman decided to begin testing nuclear weapons in Nevada beginning in 1951, which caused fallout levels to rise until just after the Test Ban Treaty was signed in 1963, must be seen as a factor in the ability of the fetus and the newborn to thrive due to impairment of the immune, hormonal, genetic, and endocrine systems. Moreover, the years of no progress (1957, 1958, 1962, and 1963) were those in which the greatest kilotonnage of bomb tests exploded over the Nevada desert, and also in the Soviet Union's test site in Chelyabinsk (now Kazakhstan). R. K. Whyte's 1992 article in the *British Medical Journal* found that there could be no other reason for the lack of improvement in first-day mortality for newborns other than fallout from atmospheric atomic bomb testing,[1] echoing the long-time contentions of Sternglass.

One way to examine whether exposure to nuclear fission products played a role in infant mortality is to look at those areas hardest hit by fallout in those years, from bomb testing and nuclear weapons and energy produc-tion. The experience of populations living near the three oldest nuclear sites is discussed below.

Oak Ridge

The nuclear era just preceded the post-war Baby Boom with the operation of three plants involved in creating the atomic bombs used in World War II. The first of these was Oak Ridge, located in a section of eastern Tennessee farmland 20 miles west of Knoxville. Oak Ridge first began emitting fission products into the environment in November 1943 with the opening of its X-10 operation (now known as the Oak Ridge National Laboratory), fol-lowed by the startup of its Y-12 complex in January 1944. Because of the urgency of the atomic bomb construction and because the stacks at Oak Ridge's reactors were unfiltered until late 1948, releases of airborne radi-onuclides were large from the outset. The majority of these chemicals drifted into the areas to the north and northeast, following the prevailing winds, before being brought to earth through precipitation and introduced into the food chain.

The effect on local infant mortality rates was swift and immediate. Between 1943 and 1944, infant deaths rose by 114% in Anderson County, where the plant is located, and 61% in the seven counties directly to the north and northeast (under 50 miles from the plant). In contrast, the rate in the rest of Tennessee and the U.S. as a whole fell about 1%. Figures in the seven counties remained above the 1943 standard throughout the rest of the decade (until the plant's stacks became filtered for the first time), while state and national rates continued to plunge.

It is likely that the effects of Oak Ridge fission products reached much farther than the seven closest counties. However, because wind and rain patterns become more variable far from the plant, only counties under 50 miles away are analyzed, for Oak Ridge and for analysis of all subsequent locations.

Hanford

Right on the heels of the Oak Ridge startup was the commencement of operations at Hanford in eastern Washington state in January 1945. Early releases of radioactive products at Hanford were perhaps the greatest of all the plants, emitting 685,000 curies of iodine alone by 1947. From 1944 to 1945, the infant death rate in Benton County, where the reactor complex is located, soared by 121%, while the total increase for Benton and the three nearest counties to the north and east was 96%. The rest of Washington state was unchanged, while a drop of 4% was experienced across the U.S.

Los Alamos/Trinity

The third locale of the atomic war effort was at a remote outpost in northern New Mexico called Los Alamos. This location was developed by the U.S. Army in 1943, not to produce nuclear fuel as Oak Ridge and Hanford were, but to conduct research. It was the headquarters for assembling the test atomic bomb "Trinity" in July 1945. In 1945, Sandia National Laboratory, about 50 miles to the southwest, became part of the nuclear weapons research effort. In the five counties immediately downwind (north and east) of Los Alamos and Sandia, the infant death rate soared by 39% in 1945. In that year, about 12.6%, or 1 in 8 babies born in these five counties did not live to age one.

The other source of nuclear fission products in New Mexico during this time was the detonation of the Trinity bomb on July 16, 1945, the success of which allowed the U.S. government to proceed with its plans for Hiroshima and Nagasaki. Trinity was assembled at Los Alamos, but sent to a remote site well to the south near Alamogordo, close to the Mexican border. The blast took place on a tower only 100 feet above the ground, kicking up large amounts of dust. The dust mixed with radioactivity and was sent in several directions in the air before settling; the low-altitude fallout cloud drifted to the north and east.[2] The fallout from the Trinity bomb test also may have contributed to rising infant death rates near Los Alamos.

All told, 185 (36%) of the 511 infants in the 16 counties closest to Oak Ridge, Hanford, and Los Alamos who died in the first year after the three plants opened would have lived if prevailing U.S. trends had continued. The Baby Boom generation had its first victims of the nuclear era.

Other weapons plants

With an excess of Baby Boomer infants dying near the initial nuclear weapons plants in the mid- to late-1940s in the first year of operations, the other major weapons plants that opened during this time can also be analyzed in a similar manner. Between 1947 and 1953, the Atomic Energy Commission opened eight plants to perform a variety of functions in the atomic bomb program.

In counties located less than 50 miles from each of these plants to the north and east, infant death rates increased in excess of the state and national rates during the first year of operation, although excesses were not as sharp as those near Oak Ridge, Hanford, and Los Alamos. In seven of the eight (except for Idaho National) there was an excess infant mortality for the five years of plant operation (Table 5.1), similar to the three oldest sites. These findings account for 144 excess deaths (about 13% of all births in the closest downwind counties), during the first year of operation alone. Because these numbers are so large, it is probable that radioactive emissions are linked with these infant deaths. In the 43 counties near the oldest 11 sites, a total of 329 excess deaths (185+144) occurred within one year, even though these areas account for only 1% of all U.S. births. This number is a very conservative estimate of infant deaths due to nuclear operations. More excess deaths occurred after the first year of operation, and in more distant or upwind counties. In addition, these 11 only represent military reactors, and do not address the 100-plus civilian reactors that opened in the latter half of the 20th century.

Nuclear power plants

Infant mortality during the baby boom years was also affected by the opening of nuclear power plants, not just weapons facilities. In the late 1950s and early 1960s, utilities began operations at the first seven electricity-producing plants. In five of the seven locations, infant mortality in counties to the north and east (under 50 miles) along with the home counties of the reactors exceeded state and national changes (Table 5.2). The only exceptions were Shippingport, in which the local, state, and national increases were similar, and Humboldt Bay, where the local rate fell faster than the state and national figures. After five years of operation, there were excess infant deaths in all seven locations except Humboldt Bay. Out of 1386 infant deaths in these 25 local counties during the first full year of nuclear operations, 162 are in excess of national trends, or about 12% of the total. This is far short of the excesses

Table 5.1 Change in Deaths Under 1 Year;
Areas Near Eleven Nuclear Power Sites
Beginning Operations Between 1943 and 1953

Site	First Full Yr of Oper.	% Change Rate Near Plant
World War II plants		
Oak Ridge, TN	1944	+61.1
Hanford, WA	1945	+95.9
Los Alamos, NM	1945	+39.4
Post-World War II plants		
Mound, OH	1947	+13.3
Idaho National, ID	1949	+9.3
Savannah River, SC	1950	+10.5
Paducah, KY	1951	+19.7
Brookhaven, NY	1951	+1.8
Fernald, OH	1952	+7.4
Portsmouth, OH	1952	+15.3
Rocky Flats, CO	1954	+12.8

Table 5.2 Change in Deaths Under 1 Year;
Areas Near Seven Nuclear Power Sites
Beginning Operations Between 1957 and 1964

Site	First Full Yr of Oper.	% Change Rate Near Plant
Shippingport, PA	1958	+3.3
Dresden, IL	1960	+8.7
Yankee Rowe, MA	1961	+25.2
Indian Point, NY	1962	+19.5
Big Rock Point, MI	1962	+23.4
Humboldt Bay, CA	1964	−8.1
Pathfinder, SD	1965	+9.7

for Oak Ridge, Hanford, and Los Alamos, but comparable to the 13% excess for the seven nuclear weapons plants built just after World War II.

The areas most proximate to these first civilian plants reveal the most adverse patterns of infant mortality. Grundy County, Illinois, the location of the Dresden plant which opened in 1959, probably witnessed the greatest damage of any county in the nation. The largely rural areas and small towns making up the county (1960 population 22,350) were exposed in the 1960s to the heaviest radioactive emissions of all civilian nuclear plants. The county averaged over 12 infant deaths a year from 1959 to 1970, but only 5 thereafter, despite a constant number of about 500 births a year.

An even closer look at effects on those living closest to nuclear power plants can be taken by studying trends in Peekskill, New York, the small town (1960 population 18,737) situated only about 2 miles from the Indian Point nuclear facility which opened in 1962. While most of the press coverage in the early years of the plant's operation centered on the thousands of Hudson River fish that were killed from large releases of hot water from the plant, babies were dying at unexpectedly high rates. Peekskill's infant deaths from 1960–62 to 1963–65 soared from 28 to 49, a rise of about 72%. After 1965, infant mortality in Peekskill quickly fell to rates at or below the national standard.

While proximity to a nuclear plant raised the chance of a baby dying, it is almost certain that fallout from atomic bomb testing in Nevada and to a lesser extent the U.S.S.R. and the Pacific played a role as well. Fallout was deposited throughout the continental U.S., while nuclear plants only existed in a handful of states. The fact that infant mortality declines halted after 1951 *in each of the 48 states* indicates that the reasons therefore are large-scale ones.

Linking infant mortality with bomb test fallout is trickier than with nuclear plant operations. While we know the location of the explosions in southern Nevada, fallout from each test traveled a different path. Moreover, fallout from each bomb went in multiple directions; often, certain radionuclides went into the upper atmosphere, while others stayed in the lower atmosphere, often in a completely different direction. Comprehensive measurements of levels of radionuclides in the diet in each part of the country were not taken from 1951 to 1957; and only monthly in nine cities from 1957 to 1960 and in 60 cities after mid-1960 were such recordings taken. Thus, it is difficult to gauge the precise "dose" to each American.

High-fallout areas

It wasn't until October 1997 that actual estimates of local dose intakes of radioactivity were made. As mentioned at the end of Chapter 4, the National Cancer Institute produced a county-by-county estimate of average iodine-131 intake for all 48 continental states from bomb tests during the 1950s. Americans ingesting the greatest amounts of iodine, a per-person average of 12 to 16 radiation absorbed doses (rads), were those living in Blaine, Custer, Gem, and Lehmi counties in Idaho, plus Meager county in Montana. These five rural counties had a total population of about 25,000 during the 1950s, plus 600 to 700 births per year. The NCI files show that most of the fallout in these counties occurred in 1952, when fallout was brought to earth through rainfall. Infant mortality rates in these counties *increased 23.1%* in these five counties between 1949 and 1951 (before Nevada testing fallout) and between 1952 and 1959 (during Nevada testing), compared to the rate for U.S. whites, which *decreased 12.2%*. American whites are used as a comparison because nearly all residents of these five counties were white.

Wet and dry areas

Another way to assess the impact of bomb test fallout on infant mortality is to compare areas by the amount of rainfall. From the limited data in the late 1950s and early 1960s (see Chapters 2 and 3), the areas with greatest precipitation generally had the highest amounts of radionuclides in the milk and water, while the driest areas had the least. The radioactive element that presents perhaps the greatest impediment to the fetus' ability to thrive is strontium. All through the early 1960s, as levels of strontium-90 peaked throughout the nation, the rainy southeast had the highest levels, and the dry southwest had the lowest. For example, in the first quarter 1964, the all-time peak year for strontium in milk, three of the four greatest readings were recorded in the southeast:

- Minot, North Dakota (58 picocuries per liter)
- New Orleans, Louisiana (54)
- Little Rock, Arkansas (43)
- Jackson, Mississippi (41)

Conversely, three of the four lowest readings during this time were cities in the southwest:

- Phoenix, Arizona (4)
- Austin, Texas (10)
- Sacramento, California (11)
- Albuquerque, New Mexico (12)

Between 1950 and 1965, the infant mortality rates for whites in Arkansas, Louisiana, and Mississippi *fell 16.6%* (from 26.52 to 22.11 per 1000 births), but conversely, white infant mortality in the dry states of Arizona and New Mexico *fell 44.9%* (42.79 to 23.56) during this time.

The gap between infant death trends in the southeast and southwest for non-whites is astounding. The rate in the three Deep South states *increased 10.8%* between 1950 and 1965 (42.29 to 46.86 per 1000 births), compared to a *decrease of 63.1%* in Arizona and New Mexico (110.49 to 40.75). The national non-white rate fell 9.4% during this time. Virtually all of the increase in the Deep South took place from 1956 to 1965, the time of the largest and most frequent bomb tests. The excess deaths of babies in Arkansas, Louisiana, and Mississippi was about 19,000 out of 93,000 infant deaths from 1951 to 1965, making it easier to understand how Sternglass calculated his figure of 375,000 unexpected infant deaths in the U.S.

Low-weight births

National patterns

Another measure of an infant's health status is birth weight. Babies born over 2500 grams, or 5 pounds 8 ounces, are generally assumed to be of "normal" birth weight (except for a tiny minority with an excessively high birth weight). Low-weight births are often premature babies born to women who deliver early (before 37 weeks gestation). Underweight newborns are much more likely to be born with congenital defects, and have a much greater risk of not surviving the first year of life.

Like infant mortality, low-weight births represent a means of assessing damage to the fetus and newborn from radiation and other exposures. Iodine exposure, for example, affects the fetal thyroid gland, which controls the fetus' growth mechanism. Strontium exposure affects the body's immune defenses, impairing not only the fetus' ability to thrive in the womb, but also affecting the mother and her ability to carry a pregnancy to term. Birth weight may actually be a more sensitive indicator of the damage due to radiation and other toxic exposures than infant mortality. Decades of superb technological advances found in neonatal intensive care units have allowed physicians to keep many infants alive whose survival would have been threatened in earlier times; thus, infant mortality rates can obscure the "true" effect of an environmental insult such as radiation. On the other hand, there is very little advanced technology can do to prevent low-weight births. The best methods for producing full-term, normal-weight babies are long-established preventive techniques that all expectant mothers should follow, such as routinely consulting a physician beginning in the first trimester, avoiding smoking and drinking, and observing proper dietary habits.

While the baby's weight has been collected for many years by hospitals, doctors, and some health departments, a number of states did not routinely require this information to be recorded by the physician or midwife. The National Center for Health Statistics did not mandate state health departments to submit birth weights for every new baby until 1948, and data on each state became available beginning in 1950. Thus, there is no comparable information on trends in birth weight for babies born before the nuclear age.

One exception is New York City, where the health department collected these statistics on a voluntary basis in 1939 and 1940 (over 98% of birth certificates contained the baby's weight). Unfortunately, the city did not produce another annual report until 1947 because of budgetary restrictions during World War II, and data in the intervening years do not exist. However, the proportion of births under 2500 grams rose from 1940 to 1947; the increase was 9.1% for whites (7.17 to 7.82 per 100 births) and 0.5% for non-whites (11.74 to 11.80). After 1947, the rate for whites moved slowly upward, while the non-white average went up sharply each year. More babies in New York City were born underweight in the post-war years than at the end of the Great Depression, suggesting the introduction of a factor like fission products into the environment caused this rate to rise.

The Baby Boom generation did not fare well in terms of their birth weight performance. While infant mortality dropped slowly between 1950 and the mid-1960s, the percentage of low-weight births actually rose. In 1950, 7.06 out of 100 white births in the U.S. were under 2500 grams, but in 1966 the figure had risen slightly to 7.22 (*up* 2%). The worst period was 1956 to 1966, which showed an increase from 6.75 to 7.22. Non-whites fared even worse;

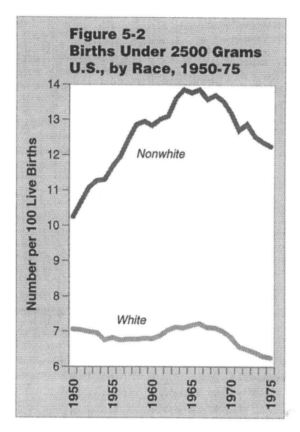

Figure 5-2
Births Under 2500 Grams
U.S., by Race, 1950-75

the 1950 rate of 10.25 had *ballooned by 35%*, to 13.86 in 1966. After 1966, rates began to decline rapidly each year for all races until the 1980s (Figure 5.2).

A broad health status measure such as the national trend of birth weight may be affected by a number of factors. Thus, looking at specific areas most affected by radionuclides (i.e., counties near nuclear weapons or power plants and those most exposed to bomb test fallout) helps to get closer to establishing a link between fission products in the diet and low-weight births.

Nuclear weapons and power plants

Unfortunately, county-specific data on birth weight is only available in *Vital Statistics of the United States* from 1952 to 1959. Thus, there is no way of knowing whether the startup of plants like Hanford and Oak Ridge affected the rate of underweight births, since we have no data from the "before" period. Nevertheless, we can look at plants that started operations during the 1952 to 1959 period to see what immediate effects occurred.

In Ohio, the Fernald plant opened in 1951, and the Portsmouth plant started operations in 1952. In the seven counties closest to and downwind of the plants (the same as used in the infant mortality analysis), the rate of births under 2500 grams fell 0.8% in 1953, while the rest of Ohio and the U.S. rose slightly. However, the local increase between 1952 and the rest of the 1950s was 8.0% (7.37 to 7.96), compared to a rise of 0.7% for the country. This excess computes to *1110* underweight babies in the seven counties, or 7% of all underweight births, in the years 1953 to 1959.

Rocky Flats Colorado, the production site of the deadly radioactive chemical plutonium, opened in 1953. In the five downwind counties closest to the reactor complex, the rate of low-weight births rose 2.1% in 1954, and 10.2% (7.99 to 8.80) between 1953 and 1954–59. This rise was far in excess of the national rate, and corresponds to an excess of *349*, or 9% of all underweight births between 1954 and 1959.

The only electric plant that opened in the 1952 to 1959 period was Shippingport. However, there was no excess found in the first year of operation, and no data exist to make a long-term projection, since county-specific numbers after 1959 were not published.

Utah and Nevada

Effects of atomic bomb testing on populations living near Nevada is the other source of radiation exposure to consider in analyzing effects on birth weight. The bomb tests in 1951 were the first in Nevada and may have represented imperfect techniques on the part of the AEC. Operators may not have been careful in initially judging the prevailing wind direction and might have failed to wait until they blew away from large urban areas such as Las Vegas and Salt Lake City before detonating bombs. Between 1950 and 1951, while the U.S. rate for whites dropped from 7.06 to 7.04, Utah's rate jumped 7.9%, from 6.93 to 7.48 per 100 births.

However, this was nothing compared to what happened in Nevada. The 1950 rate of 10.02 per 100 births soared a whopping *70.9%*, to 17.12 per 100, in 1951. This increase has never been matched by any state in the 42 years since. Nevada's rate dropped back to its customary level in 1952, but continued to remain above the U.S. average for the remainder of the 1950s and early 1960s. A revealing aspect of the incredible leap in 1951 is that virtually all of the excess occurred in the "largest" underweight babies, i.e., those weighing 2000 to 2500 grams. If in fact radiation exposure was a factor in the 1950 to 1951 Nevada change, it shows the change was relatively subtle; large numbers of babies over 5½ pounds didn't become tiny, one-pound preemies, but instead moved from the six-pound to the five-pound range. Because babies 2000 to 2500 grams are much less likely to die than very small infants, radiation exposure may not be reflected as much in infant mortality trends as it is in low-weight births. This is an illustration that birth weight may be a more sensitive indicator in showing the effects of radiation exposure than infant mortality.

High-fallout areas

Chapter 4 discusses the counties that were hardest hit by fallout containing iodine-131 and other radioactive products in the 1950s. The five Montana and Idaho counties with the greatest exposure to fallout had a 1952 to 1959 low-weight birth rate of 7.79 per 100 live births, while the 18 (mostly Montana) counties that received 9 to 12 rads per person posted a rate of 8.64 per 100. The national rate for whites during this period was 6.83 per 100; thus, about 20% of low-weight births in these high-fallout counties are excessive (*817* of 4063).

Another state-based analysis of bomb test fallout's effects involves Arkansas, Louisiana, and Mississippi versus Arizona and New Mexico (the high-fallout and low-fallout areas, respectively). Between 1950 and 1965, low-weight births in the southeast increased well in excess of national trends for all races, while falling in the southwest. The differential was especially sharp for non-whites. In Arkansas, Louisiana, and Mississippi, the rate *rocketed up 56.7%* (8.30 to 13.00 per 100), compared to a *decline of 3.5%* in Arizona and New Mexico. Nearly *2000* of the 15,000 low-weight births in the three southeast states each year in the 1950s and early 1960s can be considered to be excessive.

This finding is similar to infant mortality trends in these five states, but it appears that the gap is greater for low-weight births, again lending credence to the theory that birth weight is more sensitive to exposure to radioactive products such as strontium than infant mortality.

A final word on low-weight births concerns a related measure of infant health, that of premature births (under 37 weeks of gestation), which can also be an outcome of radiation exposure, because the cellular and genetic damage impairs the fetal ability to thrive inside the womb. National data on premature births was not kept until 1962; but since numerous studies show that many low-weight babies are born prematurely, it can be presumed that the premature birth rate also rose during the Baby Boom years.

Fetal deaths

National patterns

Another measure of infant health is fetal deaths, also known as stillbirths. Public health departments have long kept statistics on fetal deaths that only include spontaneous abortions, not abortions induced at the mother's request. Unlike infant mortality, however, the fetal death totals are not generally believed to be highly accurate; only a portion of the actual fetal deaths are reported to public health authorities, which often are not aggressive in obtaining complete totals.

Despite these drawbacks, looking at stillbirth trends is important in assessing potential health risks to infants from radiation exposure. Because radionuclides impair the body's immune system and its growth mechanism through cellular damage, exposure to either the mother, father, or child theoretically lowers the chances of the fetus' ability to survive a full-term pregnancy.

National information on fetal deaths by race has been kept only since 1945. Before 1969, all stillbirths were counted regardless of gestational age; but beginning in that year, only fetal deaths that had been in the womb for 20 weeks or longer were counted. This switch changes long-term trend analysis only slightly, as well over 80% of pre-1969 stillbirths had a gestation period of 20 weeks or longer. Public health departments often didn't bother reporting earlier fetal deaths, or never learned about them in the first place.

Again, the period 1950 to 1964 stands out as having the poorest performance in the last 50 years. From 1945 to 1950, the fetal death rate dropped by about 4% a year for both whites and non-whites, a decrease that slowed to about 1% annually during the next 14 years, the time of above-ground bomb testing in Nevada. In the period 1956 to 1964, the time of the heaviest tests and highest recorded levels of radioactive elements in the diet, there was virtually no change in rates. After 1964, fetal deaths resumed their downward trend, at an average pace of a 2% lower rate each year. Thus, despite the questions about the completeness of data, the trend observed for fetal deaths is consistent with those for infant mortality and low-weight births.

New York City

No analysis of fetal death trends in counties near nuclear facilities, or in high-fallout versus low-fallout areas will be performed here, in recognition of the highly variable reporting practices. However, analysis of New York City's rates can be undertaken, since the city began publishing these rates in 1898. From 1898 to 1943, there was virtually no change in the fetal death rate. In the period 1943 to 1964, however, the rate *doubled*, steadily rising from 70.7 to 141.3 per 1000 live births (Figure 5.3). This change means that about 90,000 more fetal deaths occurred in New York City than would be expected if the steady rate from 1898 to 1943 had continued in these years. From 1964 to 1969, with atmospheric bomb tests banned and radioactivity levels in the food chain falling, the fetal death rate dropped 13%, showing five consecutive years of decline. After 1969, a new abortion law enacted in New York State drastically cut the number of reported stillbirths, making any further trend analysis meaningless.

Congenital anomaly deaths

National patterns

Traditionally, one common cause of death in infants is congenital anomalies, or birth defects. Such defects include heart conditions, spina bifida, hydro-

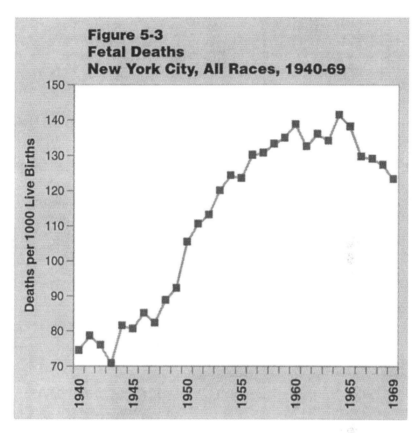

Figure 5-3
Fetal Deaths
New York City, All Races, 1940-69

cephalus, and respiratory deficiencies. As mentioned, experts agree that birth defects can be one of the adverse effects of radiation exposure. The radioactive alpha, beta, and gamma rays that directly break DNA bands and cause genetic mutations in the fetus or are passed down through an affected parent can raise the risk of an array of abnormalities at birth. Because many birth defects are not fatal and are treatable, deaths account for only a small proportion of birth defects. Still, trends in death rates among children may mirror a trend in overall incidence of defects.

Between 1950 and 1965, the congenital anomaly death rate for white American infants under 1 year decreased by 0.8% a year; just after atmospheric bomb testing ended, the decrease accelerated to 3.8% a year. Once again, non-white babies fared worse than whites. Non-white infant mortality from congenital defects increased by 1% a year from 1950 to 1965; after 1965, the rate began to steadily drop (1.8% a year from 1965 through 1969). This dramatic change before and after 1965 was likely not due to chance, but to a factor affecting the entire U.S., perhaps the drop in dietary radioactivity after the end of atmospheric bomb testing. Evidence corroborating the theory that bomb-test fallout contributed to excessively high death rates from birth defects was given by Sternglass at a conference in Berkeley, CA, in the spring of 1971. Sternglass showed that in Utah, close to and downwind from the

Nevada tests, deaths to children under five years from congenital anomalies was 77 in 1946, unchanged from 1937. Then came the bomb tests: the 1947 to 1962 average death toll in the state was 103.3 (34% higher than 1946), topped by consecutive years of 119 and 122 in 1957 and 1958, the height of the Nevada above-ground tests.[3] Radiation exposure must be seen as the culprit behind this whopping rise, argued Sternglass.

Summary

In summary, trends in infant health measures that reflect a viable immune system did not live up to expectations in the years after World War II, especially 1950 through 1965. Hundreds of thousands of unnecessary infant deaths and underweight births occurred during this time. Especially poor results are found in areas near nuclear weapons and power plants, in high-fallout areas in the Rocky Mountains, and in the high-fallout rainy parts of the southeast. Fetal deaths and mortality from birth defects also showed substandard performances during the years when atomic bombs were tested over the Nevada desert and fallout spread across the nation. Fetuses and babies, whose cell duplication in general and immune systems in particular are rapidly developing, are most susceptible to the effects of radiation exposure. Impaired cell membranes, defects in DNA/genes, and under-functioning endocrine systems affecting physical and mental growth affected each of the 75 million Boomers more than any other generation before or since, due to fallout from atmospheric nuclear weapons testing, unsafe bomb manufacturing procedures, and the crude operations of the earliest nuclear power plants. These are all relatively low-level exposures, but as the poor health record unfolded, it becomes possible to comprehend the mechanism first identified by the Canadian researcher Abram Petkau, who showed in his laboratory that *per-dose* effects from chronic low-dose radiation exposures were more severe than single, higher doses (Chapter 3).

While it appears as if nuclear fission products played a role in these negative trends, perhaps the combined effects of radiation and other toxic products hold the real key. In the 1940s and 1950s, dangerous substances such as pesticides like DDT, herbicides, industrial chemicals, plastics, and food additives either proliferated or were added to the environment. The "synergy," or combined effects of multiple toxins, is probably greater than the sum of its parts. However, radiation must be included in the list of suspected culprits affecting the infant Boomers, especially after considering the adverse trends near nuclear plants.

Despite the unmistakably negative performance in infant health across America, the trend went largely unnoticed. With the overall quality of life improving for many Americans, and with the powerful force of Cold War politics making it forbidden for any scientist to suggest that exposure to man-made radiation products had harmed thousands of children, the surviving Baby Boomers moved from infancy into childhood still seen as the healthiest group of humans in history.

References

1. Whyte, R. K., First day neonatal mortality since 1935: re-examination of the Cross hypothesis, *British Medical Journal*, February 8, 1992, 343-6.
2. Sternglass, E., *Secret Fallout: Low-Level Radiation from Hiroshima to Three Mile Island*, McGraw-Hill, New York, 1981, 87.
3. Sternglass, E., Environmental radiation and human health, in *Proceedings of the Sixth Berkeley Symposium on Mathematical Statistics and Probability*, University of California Press, Berkeley and Los Angeles, 1972.

chapter six

Health effects and the baby boomers — childhood

Health issues change as a baby moves past the first year of life and into childhood. The essential difference between infant and child health needs is the much lower death rates in children; surviving the first year means the body's circulatory, respiratory, digestive, endocrine, and other essential functions are working adequately, at least well enough to ensure survival. Some health conditions in children are more common than in infancy; these include infectious diseases such as upper respiratory, digestive, and ear/nose/throat disorders. Accidents are also more common in young children.

While the incidence of these often routine childhood diseases was not routinely tracked in the years following World War II, trends in certain diseases enable an analysis of child health. Immune-related diseases are an important indicator of child health, since the immune system affects many biological functions. These diseases also may reflect a physiological imperfection beginning in the fetal or infant period.

Cancer incidence

The most prevalent, most closely measured, and perhaps the most lethal childhood immune disease is cancer. For years, scientists have acknowledged that environmental factors raise the risk of cancer in young persons with still-developing, immature immune systems. The principle that radiation exposure increases risk of cancer in children, more so than in adults, was first asserted after studying survivors who absorbed large doses of radioactivity during the Hiroshima and Nagasaki bombings in 1945. Beginning in 1956, Alice Stewart's research extended the belief that radiation can cause cancer in children after low-dose exposures (by studying cancer rates among children whose mothers had received pelvic X-rays during pregnancy).

Historically, very little attention has been paid by U.S. public health officials to collecting data on cancer *cases* until recently. While cancer *mortality* statistics have been collected nationally for years, only one state, Connecticut, had an established tumor registry of cases in the Baby Boom years, frustrating any chance of analyzing cancer incidence for the country as a whole.

The importance of trends in cancer cases, as opposed to deaths, cannot be emphasized enough when analyzing the impact of an environmental pollutant such as radiation. If a population is exposed to high levels of radiation, the number of cancer *cases* will likely rise. However, the number of cancer *deaths* may rise less, or even fall, over time; better methods of detection plus more effective therapies keep people alive longer and even make tumors disappear. So the true effects of radiation exposure are much more accurately reflected in cancer cases, not deaths.

Thus, Connecticut's cancer registry, established in 1935, is used as a proxy for national patterns of cancer cases. Since Connecticut had no operating nuclear reactors until 1967, there is no opportunity to examine trends in childhood cancer incidence among Baby Boomers living near nuclear plants. However, Connecticut's cancer registry gives us the chance to analyze childhood cancer patterns as a result of low-level radiation exposure to distant emissions during the Baby Boom years, such as discharges from Hanford/Oak Ridge and atmospheric bomb testing in Nevada.

Childhood cancers are usually defined as those diagnosed before age 15. For each of the first five-year age groups, Connecticut Baby Boomers (born between 1945 and 1964) had higher cancer rates than did the preceding generation (born between 1930 and 1944), who had no exposure to nuclear plant or bomb test emissions at the time of their births, when the body is most susceptible to immune damage. Rates shot up *38.8%* for children age 0 to 4, *20.4%* for those 5 to 9, and *40.6%* for those 10 to 14 (Figure 6.1). However, rates for the next generation, born after bomb testing had ended (1965 to 1979) were virtually unchanged (up 3%) from Baby Boomer rates, even though this younger group ("Generation X") benefitted from better detection methods and more thorough screening than were available to the Boomers.

A total of 1893 Connecticut Baby Boomers were diagnosed with cancer as children. If the 36% rate increase by the Baby Boomers had matched the 3% rise of Generation X, then *632* fewer Connecticut children would have suffered from cancer. Since Connecticut has barely 1% of the nation's children, the total of excess childhood cancer cases among American Baby Boomers may well be in the tens of thousands. The consistent trends and the large number of cases involved give cancer patterns among Connecticut children born from 1945 to 1964 great significance.

Another clue to the radiation-cancer link among Connecticut children is to look at trends for cancers known to be sensitive to radiation. Thyroid cancer is one of these; radioactive iodine seeks out the thyroid gland, and damages the glands' ability to reproduce, thereby increasing the risk of developing this type of malignancy. For pre-Baby Boomers (born between 1930 and 1944), the disease was almost unheard of before the age of 15; only 2 cases were recorded, for a rate of 0.35 per 1,000,000 population. The Baby Boomers' total increased to 15 cases, with the rate nearly tripling, to 0.98.

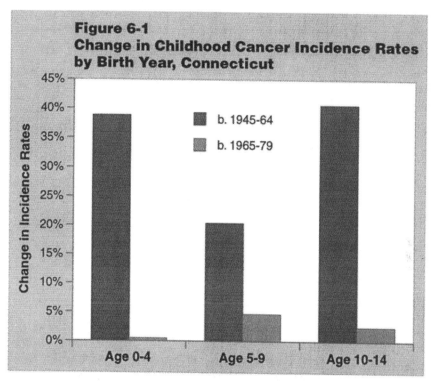

Figure 6-1
Change in Childhood Cancer Incidence Rates
by Birth Year, Connecticut

Moreover, Generation X (born between 1965 and 1979) had 23 cases, a rate of 2.29. The Boomers again did worse than their predecessors, but better than the children of Generation X, even though nobody in Generation X underwent head and neck X-rays as infants, which were found in the 1950s to carry a risk of thyroid cancer.

Another cancer sensitive to radiation is leukemia, which is much more prevalent in children than thyroid cancer, especially for those under five years of age, allowing a more meaningful analysis. The risk of developing leukemia, a cancer of the blood-forming organs, is increased after radioactive strontium and other bone-seeking chemicals routinely produced in nuclear operations and bomb tests seek out concentrates in the bone and bone marrow, where the immune system is formed and sustained. Results for leukemia (Baby Boomer rates rose 35.6%, 10.7%, and 24.0%, respectively, for children age 0–4, 5–9, and 10–14) were similar to all cancers, which is not surprising, since about one of three childhood cancers are forms of leukemia. The 29% rise for Baby Boomers under 15 was followed by a 6% increase for Generation X, or those born between 1965 and 1979 (Figure 6.2). Thus, of 633 cases of childhood leukemia among Connecticut Baby Boomers, 23%, or 146, were excess.

For children age 0 to 4, the leukemia incidence rate rose steadily as the baby boom years progressed. The highest rate, 7.93 per 100,000, was recorded in 1960 to 1964, a period when children under 5 had been born during the height of the atmospheric bomb tests in Nevada (Table 6.1). This rate, based

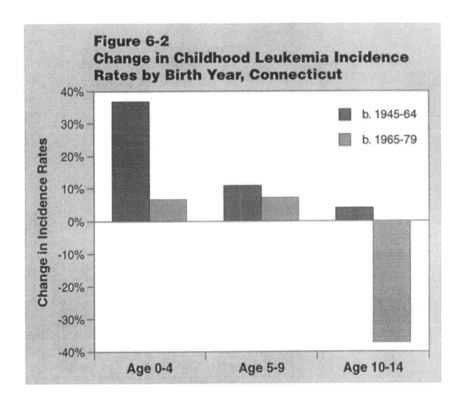

Figure 6-2
Change in Childhood Leukemia Incidence Rates by Birth Year, Connecticut

Table 6.1   Leukemia Incidence Rates,
Children Age 0-4,
Connecticut, 1935–69

Year diagnosed	Rate/100,000
1935–39	3.27
1940–44	5.25
1945–49	4.89
1950–54	6.33
1955–59	6.12
1960–64	7.93
1965–69	4.48

on 111 cases, was nearly triple that of the late 1930s. After the Test Ban Treaty went into effect, childhood leukemia incidence reverted to a lower level, rising again in the 1970s and late 1980s.

Although no state has a cancer registry as old as Connecticut's, the next oldest (Utah) reflects similar trends in cancer incidence among children. In the period 1966 to 1970, when most children were those born during the atmospheric bomb test years, cancer incidence for age 0 to 14 was 13.1 per 100,000; but from 1971 to 1990, the rate only increased to 13.2, up 0.8%. The similar rates between the Baby Boomers and Generation Xers in Utah duplicates the finding in Connecticut. For acute lymphocytic leukemia, which is the type of leukemia most sensitive to radiation exposure, the 1966 to 1970 rate of 3.4 per 100,000 children age 0 to 14 actually declined in 1971 to 1990 to 3.2. Previous research shows that thyroid cancer in young persons living in Utah quadrupled during the atmospheric bomb test years.[1] Thus, it appears that childhood cancers among Utah's Baby Boomers followed the same pattern as their counterparts living in Connecticut.

Rising childhood cancer and leukemia among Baby Boomers present compelling evidence that post-war babies were succumbing to immune-related diseases in numbers that were thousands more than expected. More than the general measures of infant mortality and underweight births, here was hard evidence that radiation-sensitive diseases were climbing among those who were exposed *in utero*, infancy, or early childhood, when the immune system is most sensitive to radiation-induced damage. Although it is unfortunate that only Connecticut tabulated cancer cases in the Baby Boom years, the large numbers of cases involved makes the increases significant. We can be reasonably certain, then, that what happened to Connecticut's Baby Boomers occurred across America.

Cancer mortality

From bomb testing

Cancer deaths are not as revealing in judging radiation's effects as cancer cases, but analyzing mortality data is a critical exercise. Many children with cancer don't live very long; this was particularly true in the 1950s and 1960s, before the development of today's more sophisticated surgery, radiation, and chemotherapy treatments. Moreover, it is important to remember that these cancer cases aren't just illnesses, but killers of little children.

Cancer mortality can be examined to detect any unusual trends among Baby Boomers, compared to the generation born before them. Not only are figures for the entire U.S. available, but specific types of cancer can be analyzed as well. The large number of cases involved allows us to break the data into children born in five- and ten-year intervals, whereas the limited number of cases in Connecticut restricted the analysis to 15- or 20-year groupings. Leukemia and all other types of cancer are examined separately (Figures 6.3 and 6.4).

Childhood cancer rates for Baby Boomers were clearly higher than those in the previous generation. Rates for persons born in the 1950s were 25 to 50% higher than those born in the 1930s, both for leukemia and for all other cancers combined. These soaring rates formed the core of why Pauling, Sakharov, and others spoke out for an end to above-ground bomb testing in the 1950s and early 1960s. The end of such testing, combined with improved treatment methods such as chemotherapy, radiation, and surgery, helped bring down cancer death rates for post-Boomers. But the damage had already

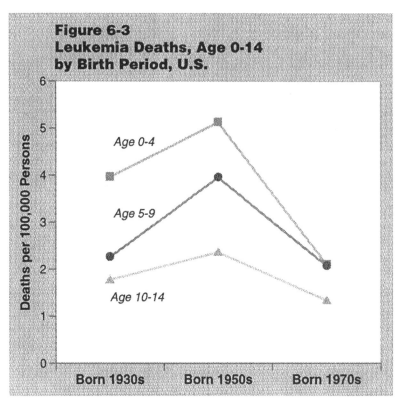

Figure 6-3
Leukemia Deaths, Age 0-14
by Birth Period, U.S.

been done, with many thousands of casualties left behind. The number of 1990 deaths in the U.S. age 0 to 14 from cancers and leukemia (1697) paled in comparison to the high of 4615 deaths in 1961. If cancer death rates for just the Baby Boomers born in the 1950s (about half of all Boomers) had been equal to those born in the 1930s, about *5000* childhood cancer deaths would have been avoided.

Near nuclear power plants

In addition to suffering from bomb test fallout, children living near nuclear plants were exposed to radiation during the Baby Boom years. The 1990 National Cancer Institute study provides some data on childhood cancer mortality trends among these Boomers. One means of assessing any effects of proximity to nuclear emissions is to take those plants that opened during the Baby Boom years, and compare cancer death rates before and after the plants were opened. Since the NCI study begins with 1950 data, the six electric power plants that opened between 1957 and 1962 were selected for study, for children under age 10. These sites include Shippingport (near Pittsburgh, PA), Dresden (near Chicago, IL), Yankee Rowe (in western Massachusetts), Big Rock Point (in northern Michigan), Hallam (in eastern Nebraska), and Indian Point (near New York City). The NCI included the

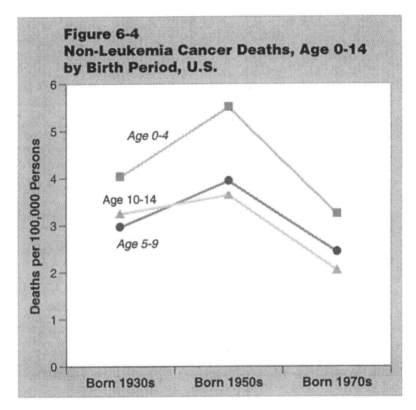

Figure 6-4
Non-Leukemia Cancer Deaths, Age 0-14
by Birth Period, U.S.

12 closest counties to the six sites, which had a total of about 2 million residents in the mid-1960s.

Mortality rates for deaths between 1950 and the plant's startup were contrasted with the period 6 to 10 years after startup; as researchers such as Alice Stewart have pointed out, childhood cancer deaths do not ordinarily occur right after radiation exposure, but often after a lag of five or more years.

Once again, the results are startling. The leukemia death rate for children under 10 in the 12 counties was 2% above the U.S. rate; but 6 to 10 years after the plants began operating, the rate leaped to 40% *above* the national mark. The same development occurred in deaths from non-leukemia cancers, moving from 9% below to 32% *above* the national average. Each of the six sites reported increases for both leukemia and cancer. The pattern near nuclear plants is consistent with the principle first laid down by Stewart of excess leukemia/cancer mortality occurring in children within 10 years after exposure to low-dose pelvic X-rays during pregnancy.

Unlike the uproar over bomb testing, there was little opposition to the production of these first-generation nuclear reactors. They were seen as a peaceful and constructive use of the atom, but the above data demonstrate they were harmful, at least to young Baby Boomers. It is difficult to precisely quantify how many excess deaths from childhood cancer occurred near these oldest nuclear plants, but about 70 of the 172 cancer and leukemia deaths in a five-year period in just the 12 counties closest to where the reactor is located may be considered excessive. Extending the analysis to later periods in these six areas, to children over nine, and to the other 100 reactors around the U.S. will likely make this excess figure grow considerably.

In Chapter 5, the southeast states of Arkansas, Louisiana, and Mississippi (which had high fallout levels in water and milk during the atmospheric bomb test years) were compared to the southwest states of Arizona and New Mexico (which had comparatively little fallout). However, since state-based cancer mortality data by age are only available since 1949, there is no way to calculate rates for pre-Baby Boomers, and any test of the association between bomb test fallout and changes in childhood cancer death rates is not possible.

Deaths from other immune-related diseases

Septicemia

One disease depending strongly on the strength of the immune response worthy of inclusion in an analysis of radiation's effects is septicemia. Commonly known as blood poisoning, septicemia (or sepsis) is a condition resulting from an invasion of bacteria and other microorganisms into the bloodstream, producing symptoms such as chills, fever, and exhaustion. Septicemia is often treatable but can be fatal when invading agents overpower the body's immune defenses. Fatalities are most common among the very elderly but can occur at any age.

Traditionally, about 1 of 100,000 American children age 0 to 14, many of them infants and babies, die each year from septicemia. Thanks to improvements in treatment, such as the development of antibiotics, doctors began to make progress in controlling the septicemia death rate in young people. From 1940 to 1945, the rate fell by nearly half, from 1.16 to 0.69 per 100,000. However, after 1945, the rate began to rise for the next 15 years; the 1960 rate of 1.07 was *58% higher* than in 1945, nearly rubbing out any progress made since 1940. At the peak of 1960, about 600 American children died of septicemia each year, of whom about 220 were excess deaths if the 1945 rate had gone unchanged. Thus, several thousand Baby Boomers died unexpectedly as children from septicemia. Rates began falling after 1960, until leveling off again in the 1980s.

Congenital anomalies

Congenital anomalies, or birth defects, have long been acknowledged to be a result of radiation exposure, due to the damage inflicted on the DNA strands by alpha, beta, and gamma rays and particles. As mentioned in the last chapter, no accurate count of birth defects in the U.S. was made until 1989. However, a national count of deaths due to congenital malformations have been maintained since the 1930s; and although many children born with one of these afflictions survive for many years, trends in deaths may provide a sampling of what the overall patterns really are.

In 1940, about 450 American children age 5 to 14 died of a congenital anomaly each year. By 1950, when children at these ages were still born before the atomic age, rates had risen 3.5% and 28.2% for ages 5 to 9 and 10 to 14, respectively. However, over the next 10 years, Baby Boomers began to fill the ranks of these groups. By 1960, rates had soared *53.9% and 52.0%* beyond the 1950 rates, meaning about *1600* excess deaths from birth defects in young children had been incurred during the 1950s. After 1960, rates dropped, reaching 1940 levels by 1970. Thus, the brunt of the excess deaths was borne by the older Baby Boomers, born between 1946 and 1955.

Pneumonia

Although pneumonia is much more harmful to infants than to children, it still is a common childhood disease that can also prove fatal in some cases, when the immune response is overpowered by the viruses or bacteria causing the infection. Before the advent of antibiotics and other drugs, this respiratory disease was feared by many parents of young children. The Spanish flu pandemic of 1918, for example, killed 550,000 Americans in a matter of weeks, many of them youths. Although this was a flu epidemic, many of the deaths were due to pneumonia which complicated some of the cases, and not the flu itself, which often ran its course in a matter of days.

In the 1940s, the introduction of new drugs induced pneumonia deaths to drop rapidly. The 1940 total of over 1500 deaths a year for children 5 to 14 slid to only 650 by 1950, representing a drop of about 60%. However, the rate only fell 20% per decade during the 1950s and the 1960s, when Baby Boomers occupied the age group 5 to 14. We would expect the decline to be greatest in the 1940s because of the new drugs; however, the under-achievement of the Baby Boomers is unsettling when coupled with similarly poor performances in infant deaths, low-weight births, cancer, septicemia, and congenital anomalies. Again, the question of whether the immune response of these young children was impaired must be raised.

Communicable childhood diseases

Cancer, while often deadly, is a relatively rare condition among children. Congenital anomalies, septicemia, and severe cases of pneumonia also have a low incidence in the early years of life. Much more common, especially a generation or two ago, is a series of viral and bacterial diseases such as measles, mumps, and chicken pox. These conditions are rarely fatal and usually resolve themselves without any treatment beyond bed rest. Whether a child develops one of these diseases, many of which are highly contagious, is at least partly dependent on the child's immune response. Again, if radiation exposure adversely affected the immune system of the young Baby Boomers, an inordinately high prevalence of these diseases may be expected.

Annual numbers of cases of conditions deemed notifiable by the federal government for each year since 1945 are available from the U.S. Centers for Disease Control and Prevention (CDC). While there are several dozen of these illnesses, only a few afflict mostly children. These conditions are aseptic meningitis, encephalitis, measles, meningococcal disease, mumps, shigellosis, streptococcal sore throat/scarlet fever, pertussis (whooping cough), and varicella (chicken pox). For some diseases, cases have been reported every year since 1945, while reporting was discontinued for some and started for others after 1945 (Table 6.2).

When the Baby Boom generation were young children, the most common childhood diseases it had to face were chicken pox, measles, mumps, scarlet fever, and whooping cough. From the late 1940s to the late 1970s, whooping cough cases fell rapidly, all but eliminating the disease. The development of a vaccine now routinely administered to infants turned this once-dreaded bacterial infection, *Hemophilus pertussis*, into a rare disorder. Even if a child develops the disease, it is readily treated by antibiotics and human serum, plus bed rest and nutrition.

The other four common childhood diseases, however, raise troubling issues for the Baby Boomers.

Measles, or rubeola, is a viral disease marked by a rash, fever, nasal discharge, and redness of the eyes. It is highly contagious, spread through droplets from the nose, throat, and mouth. In 1994, 76% of measles cases were persons under 20. Although by far the most common notifiable disease in children during the post-war years, measles was nearly eliminated after the introduction of a vaccine in 1963. However, the number of cases rose 25% from the late 1940s to the late 1950s, with the big jump beginning in 1948. The most cases in a single year occurred in 1958, when 763,094 persons were infected.

Scarlet fever is a disease caused by the bacteria *streptococci* that manifests itself through a sore throat, headache, fever, flushed face, red spots in the mouth, and a raw-textured tongue. It is also highly communicable, passed on through droplet spray or contaminated articles, food, and milk. While mild cases can be resolved through bed rest, some cases can be severe and lead to complications such as encephalitis (an inflammation of the brain). Most cases of scarlet fever are children between 2 and 8. From 1945 to 1950, the number of reported cases in the U.S. dropped from 185,570 to 64,494. In the 1950's, however, the decline suddenly reversed into a sharp rise, and by the late 1960s, the number had soared to over 400,000 a year. Although the greatest number of reported cases occurred in 1969 (450,008), mandatory reporting of the disorder ceased (inexplicably) the next year. Perhaps more than any other childhood disease, the pattern of scarlet fever is statistically linked to increasing radioactivity in the environment in the 1950s and early 1960s.

Mumps is also a highly contagious viral disease, affecting mostly children 5 to 15, that features pain, swollen salivary glands, fever, and difficulty in swallowing. Soon after the measles vaccine was introduced in the early 1960s, a similar measure for mumps was introduced, and cases dropped by

Table 6.2   Number of Reported Cases, Selected Notifiable Diseases, U.S., 1945-74

Period	Encephalitis	Measles	Meningococcal Dis.	Mumps	Shigellosis	Scarlet fever	Chicken pox	Whooping cough
1945-9	3,931	2,268,526	24,216		129,148	583,191		544,363
1950-4	8,711	2,664,185	22,349		109,158	543,042		332,450
1955-9	11,949	2,823,147	13,642		58,789	1,149,679		194,966
1960-4	10,678	2,190,391	11,937		63,494	1,713,887		74,166
1965-9	8,715	576,802	14,156	243,127 (2 yr)	60,515	2,161,292		35,329
1970-4	6,940	203,700	8,814	432,847	95,437	433,405 (1 yr)	488,536 (1 yr)	14,733

about 95% by the 1980s. Mumps was not made a notifiable disease until 1968, limiting any analysis of the Baby Boomer's experience. The first four years of national reporting (1968 to 1971), a time when most mumps victims were Baby Boomers, shows the trend in annual numbers of cases was unpredictable (152,209, 90,918, 104,953, and 124,939).

Chicken pox, or varicella, is caused by a herpes virus, and is marked by a rash of blisters, fever, and headache. In 1994, 80% of victims of the disease were under 10. Most cases have been resolved by bed rest; a safe and effective vaccine has just recently become available. Chicken pox did not become a nationally notifiable disease until 1972, making a complete analysis of Baby Boomer experience impossible. In the early 1970s, however, the disease was the most common notifiable condition in children, along with scarlet fever.

One other disease that should be mentioned is *encephalitis*, a viral inflammation of the brain and spinal cord. Encephalitis, while not as common as measles, mumps, and scarlet fever in the Baby Boom years, still raises a troubling issue. Incidence of the disease, which can result from complications of the flu, scarlet fever, measles, and other conditions, nearly tripled from 3931 in the late 1940s to about 11,000 in the late 1950s and early 1960s, before a vaccine became commercially available. The greatest increase occurred between the years 1951 and 1966, the years marked by fallout from atmospheric atomic bomb tests. Although only 46% of 1994 cases are children 0 to 19, some of the dramatic rise in the 1950s and 1960s involved Baby Boomers.

Summary

Strong evidence in child health patterns implicating impaired immune function among Baby Boomers emerged during the 1950s and 1960s. The most convincing of these indicators is childhood cancer and leukemia incidence; rates for Baby Boomers soared far beyond those for children raised in the Great Depression/World War II era. While only Connecticut collected incidence data during those years, cancer and leukemia deaths among children for all 50 states also show elevated levels among Boomers. No apparent explanation exists for this abrupt rise, which did *not* occur for post-Baby Boom children. The case for radiation causing this unexpected pattern becomes stronger after analyzing data showing Baby Boomer children living near nuclear plants were dying from cancer in the 1960s at a greater rate than children in the rest of the U.S.

Data for other diseases that depend on the immune defense also show the Baby Boomers regressing, rather than making progress. Cases of measles, scarlet fever, and encephalitis rose during the 1950s and 1960s. Childhood death rates from congenital anomalies and septicemia soared well beyond those for the pre-Baby Boomers. Pneumonia deaths fell but at a much slower rate than previously. Although these three conditions claim only several thousand children a year, their consistent pattern may be masking trends for survivors among the 75 million Baby Boomers. As the Baby Boomers moved through childhood, the number of excess victims of immune diseases, some of whom died, rocketed into the thousands.

In 1963, when many Boomers were still children, the Partial Test Ban Treaty ended atmospheric atomic bomb tests by the U.S., the Soviet Union, and Great Britain. Soon after, levels of radioactivity in the food chain began plunging. Some who earlier had feared a threat from ingesting radioactive products now relaxed, believing that the lower levels had effectively ended the threat. However, the health odyssey of the Baby Boomers does not end here. Just because this generation was becoming adolescents and adults, and because bomb tests had permanently gone underground, this didn't reverse the damage received during their fetal, infant, and early childhoood years, and continued operation of nuclear weapons and power plants posed an ongoing threat.

References

1. Weiss, E. S., et al., Surgically treated thyroid disease among young people in Utah, 1948-1962, *American Journal of Public Health*, October 1967, 1807-14.

chapter seven

Health effects and the baby boomers — adolescence and early adulthood

As the 1960s and 1970s moved along, the Baby Boom generation reached adolescence and adulthood, and began to make an impact on American life. The large contingent of babies born from 1945 to 1965 filled high schools and colleges in record numbers and quickly flooded the job market when they finished school. When 18-year-olds gained the right to vote in 1971, millions of Baby Boomers became eligible voters, drastically changing the face of American politics. Some of the group, mostly the older Boomers, served in combat during the Vietnam War or became protestors of the same war. And the first wave of Boomers got married and began producing children of their own.

The health issues faced by adolescents and young adults are quite different from those of infants and children. Cancers become more commonplace, deaths from accidents and homicide reach their highest peak, and conditions related to increased sexual activity emerge. Even though atmospheric atomic bomb tests in Nevada had ceased, adolescent/young adult health status measures can still be linked with radiation exposure. Boomers continued to be exposed in the 1960s and 1970s from the growing number of nuclear plants and from leaking underground bomb tests. Moreover, latent effects of earlier exposures can certainly manifest themselves as disease or death 15 to 20 years after initial exposure.

SAT scores/academic achievement

One issue that could reflect environmental factors such as radioactivity is mental function. Traditionally, mental capacity has been considered separately from other biological issues, but more and more, health professionals are becoming convinced that mental health reflects an organic, physiological function integrated with the rest of human biology via the brain and neurological system, and is not a separate, unrelated entity. A term achieving widespread acceptance in the past decade is psychoneuroimmunology, which is the study of the related, inseparable functions of mental, neurological, and immune system health. Malfunctioning of cells in the brain or other organs can affect humans drastically (e.g., mental retardation) or subtly (e.g., lowered intelligence).

Because radioactive chemicals damage the cell nucleus and the network of DNA strands contained therein, genetic impairment may ensue. Another major effect of radioactivity is the damage it inflicts on the immune and hormonal systems. The most obvious and immediate manifestations of this damage are birth defects, but there may also be subtle effects on biology, including brain function. One particular chemical that may figure in this impairment is radioactive iodine, which seeks out the thyroid gland and kills cells or hampers their ability to reproduce. Since this gland controls the body's physical and mental development, exposure to iodine could result in a lower level of brain function. Extreme examples include cretinism, or severe mental retardation, suffered in greater numbers in high-exposure populations, such as residents of the Marshall Islands subjected to bomb test fallout in the 1950s. However, low-dose exposure such as emissions from nearby nuclear power plants or fallout from distant atomic bomb testing may produce more subtle changes in areas such as memory, creative thinking, or concentration, resulting in a somewhat lower level of intelligence. These changes are not life-threatening, as is cancer, and may not even merit medical treatment, but if they affect millions, the negative effects of even subtle changes will be felt by society.

Establishing standard measures of intellectual aptitude and achievement is a difficult task that has challenged professional educators for years. No central body keeps track of grades in school, and even if such measurement existed, grades are subjective measures. Standard examinations such as the IQ test are given sporadically, and cannot be compared over time or in different areas. Perhaps the first chance to globally measure intelligence occurs during high school, when many students take college entrance examinations. The Scholastic Aptitude Test (SAT) has been given annually by the Educational Testing Service in Princeton, New Jersey, since the mid-20th century, and is the accepted test for college aspirants in most states. Each year, over 1 million college hopefuls take the test, consisting of verbal and mathematical aptitude questions, creating a large database for measuring trends.

In the early 1980s, Ernest Sternglass made a startling discovery that disturbed many, and he drew conclusions that angered even more. He found that after 1963, the average SAT scores plunged steadily until 1980–81, after which they resumed a very slight increase, similar to the period before 1963. Sternglass charged that the decline began with the first babies born in the atomic age (those born in 1945 or 1946 would take the SAT test 17 or 18 years later, in 1963) and ended when atmospheric bomb tests were banned (babies born in 1963 or 1964 would take the test in 1981). Moreover, Sternglass found

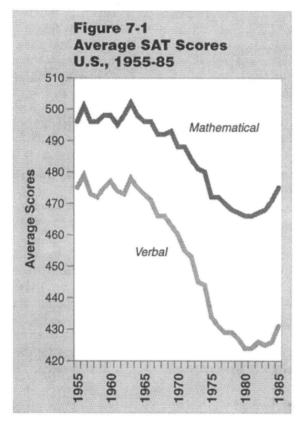

Figure 7-1
Average SAT Scores
U.S., 1955-85

that the biggest drops in average SAT scores took place 18 years after the largest bomb tests in Nevada; the biggest slump (18 points) occurred in 1975, 18 years after 1957's record-breaking year of kilotonnage released from Nevada atomic tests. The next largest drop, 11 points, occurred in 1973, exactly 18 years after the third-highest output of bombs exploded over Nevada in 1955. A storm of controversy ensued from irate critics trying to discredit Sternglass' hypothesis. While numerous alternative explanations have been offered, the matter has yet to be resolved and no one has disproved Sternglass' belief that fetal and infant exposure to fallout from atmospheric bomb testing played a major role in SAT trends.

Figure 7.1 shows that according to figures kept by the Educational Testing Service, between 1952 and 1963 the total average verbal and mathematical scores inched upward, from 970 (476 V + 494 M) to 980 (478 V + 502 M). The ensuing plunge dropped average scores by *90 points*, down to an all-time low of 890 by 1980–81. The drop was more pronounced in the verbal section (–54 points) than in the mathematical section (–36 points). In the early 1980s, the trend reversed and scores rose from 890 to 910 between 1981 and 1995. Regardless of the factor(s) behind the drastic fall, the Baby Boomers were clearly associated with a poor performance on the SAT.

One of the criticisms leveled at Sternglass was that only the college bound, and not all high school seniors take the SAT, and that between 1963 and 1980, with college education more accessible and more necessary for employment, an increasing proportion of seniors took the exam, invalidating any analysis of trends. However, scores from a separate aptitude test of *all* students, not just the college bound, bear out Sternglass' conclusions. Beginning in 1970, the U.S. Department of Education has conducted assessments of a representative sample of *all* 9-, 13-, and 17-year-old U.S. children (in public and private schools) every few years. Examinations in science, mathematics, reading, and writing are given, and are scored on a scale of 0 to 500. As scores rose in the 1980s, professional educators used the data to claim the educational system's increasing effectiveness.[1]

However, the data also show Baby Boomers scoring progressively worse and the next generation improving. In science and mathematics, scores for each of the three age groups reached a low point for children born in 1964–65, when levels of long-lived radionuclides such as strontium-90 were at their peak in the American diet. Children born in the late 1960s and 1970s scored progressively higher with each administration of the test. Results for reading were almost as striking; scores were consistently low among Baby Boomers, and rose for those born after the mid-1960s. Writing examinations were only given beginning in 1984, after the last of the Boomers had graduated from high school, so no comparisons could be made with members of Generation X.

Weight and height

After adolescence, many persons are close to the height and weight they will be during their adult life. Average height and weight in a teenage population is also an indirect measure of health status, and in a progressive society like 20th-century America, one would expect people to become taller and larger over time due to food supply, better nutritional habits, increased access to prenatal and other health services, more emphasis on physical fitness, and avoidance of certain debilitating diseases that affected earlier generations.

Numerous studies of child height and weight have been performed over the years. A classic study published in 1963 summarized many of these earlier works, and constructed trends over more than half a century. White males were the most-studied race and gender in the report, especially those 15 years of age. As expected, increases in height and weight were significant. Fifteen-year-old boys in 1960 were 5 inches taller than they were in 1892, or just under 5'7" compared to just under 5'2". Accordingly, the modern boys were also 30 pounds heavier than those from the late 19th century (133.4 versus 103.2).[2] Average weight and height had been rising steadily and showed no signs of tapering off. (Figures for each period are averages for three to four studies each, representing hundreds of white boys.)

In 1975, the U.S. Department of Health, Education, and Welfare published a nationally representative sample survey of children in 1971–72. The stunning finding of this new compilation was that *the Baby Boomers stopped growing*, an unprecedented event since data were first collected in the 1870s. From 1960 to 1972, the height of 15-year-old white boys was unchanged (just under 5'7"), and they had lost an average of two pounds (133.4 to 131.0).[3] It is critical to note that we are comparing boys born in 1945 and 1957; the former were born at the dawn of the atomic age, and the latter were born during the year of the greatest kilotonnage yielded from atmospheric nuclear tests in Nevada. The 1957 birth cohort spent their youths amidst a time of economic plenty, medical advances, greater knowledge about proper nutrition, and a marked national emphasis on physical fitness, a priority spearheaded by President Kennedy while he was in office. So the abrupt halt to growth in boys is even more astonishing. But radiation, especially the iodine that affects hormones in the fetal/infant thyroid gland and their ability to promote physical growth, may have overriden these other factors, and stopped years of steady growth in its tracks.

This chapter begins with two relatively indirect potential health outcomes of radiation exposure, mental and physical status. There is certainly an abundance of possible reasons why SAT scores dropped and physical size stagnated for Baby Boomers, and some are likely to contribute to these trends. But after a poor showing in immune-related diseases by Boomers as infants and children, especially thsoe near nuclear sites, radiation exposure must be regarded as a potential contributor to these unexpectedly negative trends.

White blood cells

Another indicator of health that may represent subtle damage to masses of people rather than the presence of disease in just a few is concentration of leukocytes, or white blood cells (WBCs). These cells form the immune system's "army" of defense against infection from invading agents such as bacteria and viruses. Granulocytes, which account for 70% of all white blood cells, are formed in the bone marrow. The other two types of white blood cells are lymphocytes and monocytes, formed in the lymphoid tissue.

One of the simplest and most commonly performed laboratory tests in the U.S. medical system is a count of white blood cell concentrations in blood, which can inform physicians of the strength of a patient's immune response or possibly suggest presence of a disease. While the medical literature is full of articles describing blood counts in persons with certain conditions, very little information exists on white blood counts for the general population. The National Center for Health Statistics survey of WBC concentrations in the U.S. population in 1971 through 1975 went virtually unnoticed, a routine set of data destined to sit anonymously on dusty shelves in libraries (Table 7.1).

However, results of the survey are striking. For both whites and blacks, average white blood cell counts were lower for persons age 6 to 17 (born mostly between 1956 and 1967) than for middle-aged adults age 25 to 54 (born between 1919 and 1948), even though the children's counts normally should be slightly higher.[4] The period 1956–67 corresponds to the period in which the American diet contained its highest radioactivity levels.

Another, perhaps more ominous finding concern data on persons with the lowest WBC counts, or those most vulnerable to developing disease and least able to defend against it. Of men born from 1956 to 1967, about 9% had under 5 billion white cells per liter of blood, compared to only about 5% for those born between 1919- and 1948, while Baby Boomer women had a percentage only slightly higher than their older counterparts. Five billion WBCs per liter of blood represents a rough lower limit for a "normal" level of leukocytes and is used here as a cutoff point.[4]

The majority of Baby Boomers had white cell counts in the normal range, and differences between older and younger persons may seem modest at first glance. However, these figures are enormously important; even subtle differences may decide whether a person does or doesn't contract a disease. Raising a generation's susceptibility to immune diseases even slightly could cause thousands, even millions of excess illnesses and deaths, robbing society of its productive members and costing enormous amounts in medical treatment. Higher rates of the flu, the common cold, and allergies would have a negative impact; but more cancer, pneumonia, and AIDS would have a crushing impact. This topic is addressed in more detail in Chapter 8.

Cancer incidence

As seen in the previous chapter, one of the most important measures of immune functions in a population is cancer incidence. As the Baby Boomers grew into adolescents and adults, cancer began appearing in much greater numbers than in their parents' generation, duplicating the pattern demonstrated in Chapter 6. Using figures from the Connecticut tumor registry, overall cancer rates for persons age 15 to 19 rocketed up 55.4% for persons born between 1945 and 1964, compared to those born 20 years earlier. The rate for Generation X (born between 1965 and 1984) continued rising, but at a much slower pace (19.5%). The same pattern occurred for persons age 20 to 24; the Baby Boomer jump of 43.4% was followed by an increase of only 4.8% (Figure 7.2).

Of 2378 cancers diagnosed in Connecticut Baby Boomers between age 15 and 24, 767 fewer would have occurred if rates had remained unchanged. If Connecticut is any sort of a barometer of the United States, it is likely that tens of thousands of excess cancer cases occurred in Boomers age 15 to 24 across the nation. Radiation exposure in infancy and childhood can lead to increased cancer risk in teenage and early adult years, especially those sensitive to radiation. Thyroid cancer, the risk of which is raised after exposure to radioactive iodine, was virtually unheard of at age 15 to 24 for the generation born before 1945 (about 2 cases per year in Connecticut). However, this number jumped to about 9 cases per year among Baby Boomers, more than doubling the rate. The generation born after 1965 experienced slightly less thyroid cancer from age 15 to 24 than their Baby Boomer counterparts.

Table 7.1   White Blood Cell Concentrations by Age and Sex,
U.S., 1971–75

	Men			
	Average WBCs*		% with WBC <5.0	
Birth Yr	White	Black	White	Black
Post-Boomers				
1968–72	8.5	8.1	4.4	4.4
Baby Boomers				
1962–67	7.6	7.1	6.4	15.8
1956–61	7.2	6.6	9.6	23.2
1949–55	7.3	7.0	5.5	12.0
Pre-Boomers				
1939–48	7.7	6.6	4.8	14.3
1929–38	8.0	7.0	3.3	16.4
1919–28	7.8	7.1	3.7	10.0
1909–18	7.6	5.8	6.1	31.0
1899–1908	7.5	6.4	4.7	26.4

	Women			
	Average WBCs*		% with WBC <5.0	
Birth Yr	White	Black	White	Black
Post-Boomers				
1968–72	8.5	8.5	2.7	5.4
Baby Boomers				
1962–67	7.4	7.0	6.8	18.0
1956–61	7.6	6.8	5.8	17.9
1949–55	7.9	7.4	3.4	17.1
Pre-Boomers				
1939–48	7.7	6.9	6.1	18.6
1929–38	7.8	7.3	6.3	11.9
1919–28	7.5	6.7	5.9	11.6
1909–18	7.3	6.9	6.3	23.2
1899–1908	7.2	6.9	7.9	22.6

* Billion white blood cells per liter of blood

Increases in leukemia rates for Boomers age 15 to 19 and 20 to 24 were only
7.2% and 10.8%, respectively, even though leukemia risk is raised by exposure to bone-seeking radioactive products such as strontium.

One at-risk subset of the Connecticut population is that living near the two
nuclear power plants in the state, Haddam Neck and Millstone, which began
operations in 1967 and 1970, respectively. Persons who were teenagers living
in the home counties of the two sites when they opened thus received a "double
dip" of radiation exposure early in life; as part of the Baby Boom generation,
they were exposed to distant fallout from weapons plants and bomb tests in
infancy and childhood, plus the routine emissions from local reactors in ado-

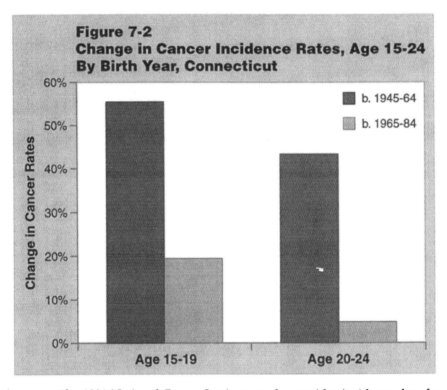

Figure 7-2
Change in Cancer Incidence Rates, Age 15-24
By Birth Year, Connecticut

lescence. The 1990 National Cancer Institute study provides incidence data for the home counties of the two reactors (Middlesex County, where Haddam Neck is situated, and New London County, the site of Millstone).

Between 1950 and 1962 in Middlesex County, cancer cases in persons 10 to 19 (pre-Boomers) was 35% below the state rate. In 1968–72, with Haddam Neck operating and Baby Boomers making up this age group, the county rate rocketed to a level 14% *above* the Connecticut standard. The same pattern occurred in New London County; the pre-Boomers age 10 to 19 developed cancer at a rate 26% below the state average, only to witness the same rate for the Boomers advance to 14% below the state average in the 10 years after Millstone started operations. Despite the small number of cases (72 in the two counties over 10 years) limiting the significance of the trend, this "double dip" of exposure early in life should be considered seriously, and will be examined more thoroughly in the next chapter.

Cancer mortality

As seen in the last chapter, by the time most Baby Boomers reached their teenage years, medical science had made significant advances in treating cancer in young persons. Cancer mortality rates plunged for all races, ages, and sexes, making analysis of U.S. trends more reflective of technology changes than of any effects of environmental threats such as radiation. However, it is important to examine whether or not the "double dip" group, i.e., those who were exposed to bomb test fallout *and* nuclear power plant emissions in their youth, fared as well as the rest of the country.

Data from the NCI study illustrates the difference in cancer mortality for persons 10 to 19 living near nuclear facilities. Counties with some of the oldest and most contaminated weapons plants had rates well above the national average during the 1960s, such as Hanford (64% above U.S. rates in 1964 through 1968), Oak Ridge (17% higher in 1964 through 1968), and Savannah River (47% higher in 1961 through 1965).

Data allowing a true comparison of cancer death rates for age 10 to 19 between pre-Baby Boomers and Baby Boomers are available for the 12 counties nearest the six oldest nuclear power plants. In the years before the plants began operating, local cancer death rates were 5% above the U.S. average; but 6 to 10 years after reactor startup, the rate was 26% *higher*. Leukemia deaths also rose, from 26% above to 35% above the national standard. Again, these six areas maybe just be the "tip of the iceberg" of health effects near nuclear plants; damage may have occurred well beyond the two counties nearest each site, may have occurred at newer plants, and may have affected persons in age groups other than the Baby Boomers.

Mortality, all internal causes

In the 1950s and 1960s, public health officials took note of a steady rise in death rates among Americans in their teens and early 20s, following many years of steady decline. They were quick to point out that the reversal was due to more accidents resulting in deaths of young people. In the post-war years, many more Americans began to drive cars, including younger drivers who are at greatest risk for accidents. The construction of the interstate highway system beginning in the Truman administration gave young Americans more chances to drive at high speeds, further raising the death toll.

While deaths from accidents rose, they do not completely explain the greater death rates among teenagers and young adults in the late 1950s and 1960s, a fact missed by many researchers. Subtracting accidents, suicide, and homicide from the total number of deaths leaves deaths from "internal," or medical causes. This statistic is still a broad one, made up of a variety of diseases such as heart disease, cancer, and infectious diseases. However, it is a better way of reflecting potential effects of an environmental factor such as radiation exposure on health, since accidents, homicide, and suicide are not directly affected by toxic substances.

Despite the fact that death rates among teens and young adults from internal causes are low, an unusual pattern began to unfold beginning in the mid-1950s. For persons age 15 to 19, the rate plummeted from 79.0 to 34.4 per 100,000 between 1946 and 1956. However, from 1956 to 1971, the number barely moved, dipping only from 34.4 to 30.4, before resuming its downward trend. From 1956 to 1971, persons age 15 to 19 were almost all born in the

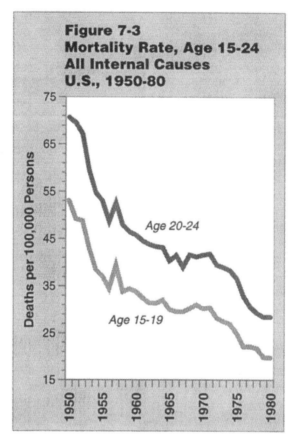

**Figure 7-3
Mortality Rate, Age 15-24
All Internal Causes
U.S., 1950-80**

early 1940s, late 1940s, and early 1950s, the oldest of the Baby Boomers. The younger Boomers fared much better, as the 1971 to 1981 rate fell from 30.4 to 19.0 (Figure 7.3).

A similar pattern occurred for the 20 to 24 year age group. From 1946 to 1962, the internal death rate slid from 119.2 to 43.8; but from 1962 to 1974, the rate changed little, from 43.8 to 38.3, before resuming a more typical decline. Once again, the population most affected here are those born between 1940 and 1954. The abrupt halt in the dropping death rates for the 15 to 24 age group came at a time when medical technology was improving to save more lives, when numbers of hospitals and doctors were growing, and when public health programs like Medicaid made access to care a reality for millions, raising the possibility that one or more environmental factors account for the lack of progress.

Pneumonia

As mentioned in Chapter 6, pneumonia deaths in the U.S. fell sharply after the introduction of antibiotic drugs in the 1940s. This pattern also occurred among adolescents and young adults; between 1940 and 1950, annual pneumonia deaths in the U.S. to persons age 15 to 24 fell from 2100 to 600.

The rate continued to fall in the 1950s. But in the next decade, when the oldest Boomers reached adolescence, the rate stagnated, before falling further after 1970. The rate of pneumonia deaths among those age 15 to 24 for persons born from 1946 to 1955 was just 3.1% lower than those born a decade earlier, bringing the years of outstanding progress to an abrupt halt. Again, the oldest Boomers appear to have posted the worst record of any age group in the past half century, as they had for a number of other diseases. The 15 to 24 pneumonia death rate for Boomers between born 1956 and 1965 fell 66% below that of the older Boomers, resuming the previous decline.

Sexual diseases and reproductive disorders

Gonorrhea

As the Baby Boom generation grew into adulthood, it became sexually active and began reproducing. These activities bring with them a new set of health concerns, such as venereal disease, fertility problems, infections of the reproductive organs, and conditions related to pregnancy and childbearing. Sexual and maternity problems are not often discussed as a result of environmental insults, such as radiation. Cancer, for example, is much more directly linked to radiation's toxicity because of the disruption of cell reproduction after radioactive products harm the cell membrane. However, sexual and maternity disorders are dependent on the strength of the body's immune defenses, plus the balance of its hormonal system. Radioactivity is known to be harmful to both, and thus it is important to examine trend data for Baby Boomers.

Several sexually transmitted diseases must be reported to public health departments, making trend data available. One of these indicators is gonorrhea, a disease marked by inflammation of the genital organs and urethra, and contracted through sexual activity. Despite the discomfort of gonorrhea, virtually all cases are cured by penicillin or other antibiotics, first made available in the 1940s. Currently, about 91% of gonorrhea cases occur in persons 15 to 39 years old, according to the U.S. Centers for Disease Control and Prevention, in its October 6, 1995 issue of the *Morbidity and Mortality Weekly Report*. After World War II ended, gonorrhea was on the decline, with 1957 cases (214,496) barely half of the 1947 total (380,666). Thereafter, a sharp increase in reported cases took place, reaching over a million in each year from 1976 to 1980. The number of cases declined steadily since then, falling to 413,647 in 1994 (Figure 7.4). Changes in sexual practices influence the trend in reported cases, and the sexual "revolution" beginning in the 1960s helped to boost the incidence of the disease. However, another factor of whether a person exposed to gonorrhea develops the disease is the ability to fight off the offending organism *Neisseria gonorrheae*.

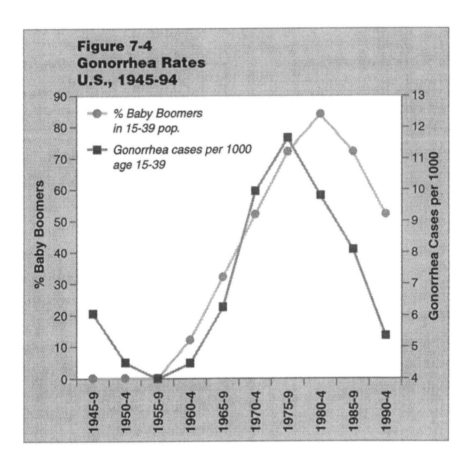

Figure 7-4
Gonorrhea Rates
U.S., 1945-94

Syphilis

Syphilis is another notifiable disease representing a sexually transmitted organism (the spirochete *Treponema pallidum*). Syphilis is also curable with penicillin if the medication is administered in the primary stage (usually a genital sore) or secondary stage (skin eruption and inflammation of several organs). If allowed to progress to a later stage, syphilis can damage the heart and central nervous system, and may even cause blindness. In 1994, about 81% of primary and secondary syphilis cases were reported in persons 15 to 39 years old, according to the CDC.

The number of primary and secondary cases reported after World War II dropped sharply, from 94,957 in 1946 to 7,147 in 1954. Beginning in 1960, when Baby Boomers entered the 15 to 39 age group, incidence tripled to just over 20,000 per year by the early 1960s. However, unlike gonorrhea, syphilis rates remained fairly constant through the 1960s and 1970s, and did not fall after 1980, even though Baby Boomers were aging out of the 15 to 39 year age group (Table 7.2).

Table 7.2 Reported Cases of Syphilis
(Primary and Secondary), U.S. 1945–94

Year	Cases
1945–9	374,625
1950–4	64,657
1955–9	36,397
1960–4	102,283
1965–9	103,954
1970–4	120,404
1975–9	116,221
1980–4	153,388
1985–9	174,818
1990–4	174,256

Infertility

Another reproductive system problem is infertility. This condition is not tracked regularly like venereal disease, but has been the subject of periodic surveys of the U.S. population. The U.S. Public Health Service conducted such surveys in 1965 and 1976, which enables a basic trend analysis to be made. The USPHS excluded any women rendered sterile from surgery in its definition of infertility. One would expect that over time, infertility rates would decline, due to improvements in living standards (such as better nutrition) and enhanced techniques devised by medical experts to improve fertility.

The surveys found that by 1976, infertility among women under 30 (Boomers) was rising, while the rate for women over 30 (pre-Boomers) was falling (Table 7.3). Some of the detailed findings are startling. For example, the percent of non-white women age 20 to 24 who were infertile rose from 3.4% to 15.4% between 1965 and 1976. The former group was born in the early 1940s, while the latter was born in the early 1950s. Even though such a survey may or may not have been repeated since then, the trend is an ominous one, as it affects not just the Boomers' well-being, but their ability to pass life on to future generations. Objective researchers should review the biological actions of a chemical such as strontium-90, which soon after ingestion turns into its daughter product yttrium-90, to measure any correlation between radioactivity and infertility. Yttrium-90 moves from its original home, the bone surface, to other parts of the body, including the reproductive organs.

Summary

The Baby Boomers' string of unexpectedly high rates of immune-related disease continued as they moved into adolescence and young adulthood. Adverse trends from childhood continued; cancer cases rose sharply, and

Table 7.3   Percent Women Infertile,
by Age and Race, U.S., 1965 and 1976

Whites	1965	1976	Change
Baby Boomers			
Age 15–19	0.6	2.0	+1.4
Age 20–24	3.4	5.6	+2.2
Age 25–29	6.1	8.4	+2.3
Pre-Baby Boomers			
Age 30–34	10.8	9.5	–1.3
Age 35–39	13.4	11.4	–2.0
Age 40–44	18.5	14.6	–3.9
Non-whites	1965	1976	Change
Baby Boomers			
Age 15–19	0.0	3.7	+3.7
Age 20–24	3.4	15.4	+12.0
Age 25–29	7.1	11.2	+4.1
Pre-Baby Boomers			
Age 30–34	15.7	18.1	+2.4
Age 35–39	24.4	23.3	–1.1
Age 40–44	39.0	28.8	–10.2

cancer mortality near nuclear plants did not improve as much as it did elsewhere in the U.S. Pneumonia deaths leveled off after many years of large decreases. Deaths from all causes (excluding accidents, homicide, and suicide) leveled off as well, compared to the large reductions enjoyed by the generations older and younger than the Baby Boomers.

In addition, some disturbing trends emerged in other areas that reflect, at least partially, immune function. SAT scores eroded for children born after World War II, compared to slight increases for the Depression babies and Generation Xers, possibly signaling a decline in mental function. Average weight and height in teenage boys stopped for persons born after 1945, after decades of steady growth. Gonorrhea cases increased from 1 million in the late 1950s to 5 million in the late 1970s. Infertility rose among Baby Boomers, while falling among women born before the atomic age. And perhaps most troubling, Baby Boomers were found to have the lowest white blood cell counts of any American age group, raising a host of questions over how much damage these compromised immune systems would be able to prevent over the course of their lifetimes. Measures like SAT scores and white blood cell counts emulate low-weight births in suggesting that impairment to the Boomers might be slight insults to the masses, and not just a matter of a few more "rarities" like childhood cancer.

The overall record was another strike against babies born from 1945 to 1964, matching the poor performance in infancy and childhood. The Boomers had amassed a growing list of excess cases of disease and death, even before moving into adulthood, when risks of contracting most diseases are much higher. Perhaps the most troubling aspect of the continued slippage in immune-related health for Boomers was the fact that, in the 1960s and 1970s, a number of toxic and carcinogenic chemicals were banned from use. During the dawn of this age of environmental awareness, products like DDT, red dye #2, and leaded paint and gasoline were taken off the market. This change should have helped reduce immune diseases, like cancer, but instead the parade of bad news for Boomers rumbled on.

References

1. *The New York Times,* December 13, 1995, B16.
2. Meredith, H., Change in the stature and body weight of North American boys during the last 80 years, in *Advances in Child Development and Behavior, Volume I,* Lipsett, L. P. and Spiker, C. C., eds., Academic Press, New York, 1963, 71-105.
3. National Center for Health Statistics, *Preliminary Findings of the First Health and Nutrition Examination Survey, United States, 1971-72: Anthropometric and Clinical Findings,* U.S. Department of Health Education and Welfare Publication (HRA) 75-1229, Washington, DC, 1975, 34-7.
4. *Textbook of Pediatrics,* 13th Edition, Behrman, R. E., Vaughan, V. C., and Nelson, W. E., Eds., W.B. Saunders, Philadelphia, 1987, 1062-5.

chapter eight

Health effects and the baby boomers — middle age

The Baby Boomers entered middle age in the 1980s and 1990s with a checkered past history of immune and other radiation-related diseases. Their earliest years had been marred by stagnant infant mortality and rising underweight births; and their childhood were and early adulthood were marked by rising cancer incidence, below-expected performance in mental (scholastic) function, more communicable disease, and growing infertility. However, no real alarm that the generation had been maimed by an environmental poison existed among professionals and the population. Fatalities from many immune-related conditions among the young were generally low when compared to the health threats that plagued earlier generations, like tuberculosis and pneumonia. Thus, these 75 million Baby Boomer Americans were still viewed as a privileged generation. The creeping monster of immune-related disease among this generation was not taken seriously, and the image of the healthiest generation in history persevered.

In the 1980s and 1990s, however, any casual attitude about immune disease in the generation born between the mid-1940s and the mid-1960s ran into a buzz saw of proportions not seen for decades, a bombshell known as AIDS.

AIDS

On June 5, 1981, the U.S. Centers for Disease Control and Prevention published an article in its *Morbidity and Mortality Weekly Report*. It focused on a report of five cases of Kaposi's sarcoma, a rare cancer, diagnosed in five homosexual men in Los Angeles. While this was an unusual event that the CDC felt was worthy of reporting, the article went virtually unnoticed.

The report, as everyone now knows, was the first official acknowledgment of the disease known as Acquired Immune Deficiency Syndrome, or AIDS. In time, researchers discovered that the condition first reared its ugly head in Central Africa in the late 1970s; affected mostly homosexual men and intravenous drug users (in the U.S.); and occurred mostly in young and middle-aged adults. Almost overnight, the disease became a worldwide pandemic, including the United States. The disorder was linked with the presence of the Human Immunodeficiency Virus (HIV) in the bloodstream, spread through blood and semen. AIDS is also marked by a steady deterioration of the number of CD-4 cells, a type of white blood cell, that rendered the human immune system progressively weaker until it succumbed to any of a variety of infectious diseases that the body couldn't counter. An infected person would often not show any signs of illness for a long time, about 10 years on average, while the CD-4 cells disappeared. After the first infection appeared (termed the onset of "full-blown AIDS"), death occured quickly; even in the late-1990s, the average time between onset of full-blown AIDS and death is slightly above two years.

The long 10-year latency between infection and illness limited the number of full-blown AIDS cases in the early 1980s. The initial sluggish response of the medical and public health communities, and continuing high-risk behavior enabled AIDS cases in America to soar in the late 1980s and 1990s. The increase was assisted somewhat by revisions of the definition of AIDS in 1987 and 1993 to count more persons as full-blown AIDS victims who were previously considered just to be HIV-positive; but even adjusting for these changes, the number of new cases continued to grow each year until 1997. By June 30, 1997, 612,078 Americans had been diagnosed with full-blown AIDS (a slight undercount due to misdiagnosis and delays in reporting cases to public health departments), according to the CDC's *HIV/AIDS Surveillance Report*. Perhaps another 500,000 Americans were positive for the HIV virus, but not yet ill with AIDS. A total of 379,258 Americans had lost their lives to AIDS by mid-1997, more than the entire American death toll during World War II.

New AIDS cases rose steadily from 1981 to 1992 (Figure 8.1). Trends after about 1991 are somewhat skewed because of the revised AIDS definition introduced in 1993; but experts generally believe the number of newly diagnosed AIDS cases has leveled off at just under 60,000 a year. Beginning in 1996, the use of new combinations of drugs, featuring those from the protease inhibitor class, offered the potential for slowing the spread of HIV in the body. However, these new drugs are not yet considered an AIDS elixir. Some patients show no benefits, some cannot tolerate the drugs, and long-term effects are not yet proven. Moreover, the medications are extremely expensive and unaffordable to many HIV-positive persons.

It is clear that much of the AIDS burden fell on those born in the two decades after World War II. About 76% of the newly diagnosed cases in 1988 were persons born between 1944 and 1963, along with 72% for new cases from 1993 to 1997. If the current trends hold, it is quite possible that full-blown AIDS cases in the U.S. will reach 1 million by the year 2003, meaning that over 700,000 Boomers will be among their ranks.

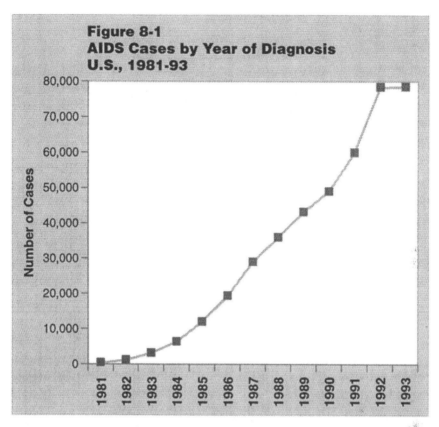

Figure 8-1
AIDS Cases by Year of Diagnosis
U.S., 1981-93

AIDS quickly helped turn around years of progress in death rates made by young American adults. From 1970 to 1983, mortality for persons age 25 to 34 (excluding accidents, suicide, and homicide) fell 31.3%, from 71.7 to 49.2 per 100,000. But after the AIDS epidemic took hold, the rate *moved up* 45.3%, to 71.5, in 1992, the same rate it had been in 1971. The same about-face occurred for those in the 35 to 44 group; a decline of 38.8% from 1970 to 1983 was replaced by a rise of 19.9% from 1983 to 1992. Rises were particularly sharp among men and non-whites. For all other age groups, rates were falling after 1983 (except for age 15 to 24, largely because of increases in homicide, suicide, and accidents).

The HIV virus is the direct cause of AIDS, not radiation exposure. However, clinical researchers are still trying to understand why some persons infected with the virus develop AIDS relatively quickly, why others do so more slowly, and why some have lived almost 20 years with the virus in good health. It is quite possible that the solution to this question lies in the ability of a body's immune system to fight off the invading agent (HIV). More research remains, but the fact that the immune-compromised Baby Boomers account for three-fourths of AIDS cases creates the possibility that radiation and other toxins may be factors in developing AIDS.

Deaths from all causes

Chernobyl and 1986

The 1986 Chernobyl disaster that spread low-level radioactive fallout across the U.S. (Chapter 4) was another factor affecting Baby Boomer health. As Gould and Sternglass showed, U.S. total deaths and infant deaths soared during the summer following the accident. In that year, death rates for all causes among young American adults jumped faster than in any year since 1918, the year of the catastrophic influenza pandemic. For persons 25 to 34 (born between 1952 and 1961, the peak years of Nevada atmospheric bomb testing), the 1986 death rate leaped 9.3% above the 1985 mark; and for those age 35 to 44 (born between 1942 and 1951), the rate increased 3.5%. The adjusted death rate for all Americans *fell* 1.2% that year. Thus, it appears that a "double hit" of radiation exposure — bomb test fallout in childhood making the immune system vulnerable and Chernobyl fallout in young adulthood inducing sick people to die faster — is linked to excess 1986 death rates among Baby Boomers.

Near nuclear reactors

Another group that may have suffered a "triple hit" of radiation exposure also shows up poorly in 1980s death rates. Persons age 25 to 44 living near nuclear weapons facilities in 1986 make up such an unlucky category. They (1) ingested Nevada bomb fallout as infants and children; (2) absorbed emissions from the local nuclear facilities throughout their lives; and (3) were exposed to Chernobyl radioactivity as young adults. Perhaps the most vulnerable of the "triple hit" groups are those living near the three weapons plants manufacturing weapons-grade fuel (Hanford, Oak Ridge, and Savannah River).

In the Augusta area (four counties near Savannah River) and the Knoxville area (seven counties near Oak Ridge), each with about 300 annual deaths age 25 to 44, mortality rates for persons age 25 to 34 and 35 to 44 rose during the late 1980s, after dropping sharply in the previous decade. Figure 8.2 reveals the extent of the turnaround; in 1990, the 25 to 34 and 35 to 44 age groups made up the younger and older Boomers, respectively. Although AIDS deaths occurred in these metropolitan areas, none are known as high-AIDS cities like New York, Newark, San Francisco, or Miami. Thus, it is likely there are reasons other than AIDS that account for this abrupt turnaround.

Cancer incidence

All cancers

As Boomers became adults beginning in the 1970s, the first national database on cancer incidence was developed. In 1973, as part of President Nixon's War on Cancer, the federal government began publishing yearly reports on national cancer incidence, based on five states and four cities representing

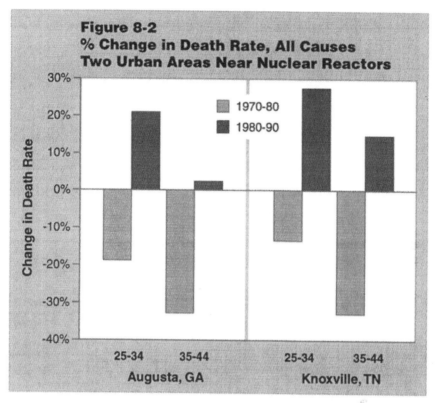

**Figure 8-2
% Change in Death Rate, All Causes
Two Urban Areas Near Nuclear Reactors**

about 9% of the U.S. population. The data, compiled by the National Cancer Institute, provides a fair estimate for incidence in the 50 states.

This annual report series serves as the basis for one of the more startling findings in cancer trends since the 1970s. Incidence of cancer in the 35 to 44 age group fell 5.1% from 1973 to 1979; however, in 1979 an abrupt reversal took hold, and a steady rise occurred throughout the 1980s. After 1979, as Baby Boomers were added to the age group, the rate *climbed 13.7%* from 1979 to 1988 (152.3 to 173.1 per 100,000), and remained steady from 1988 to 1992 (Figure 8.3). Increases from 1979 to 1988 occurred for white males (25.3%), white females (8.7%), black males (24.8%), and black females (8.2%). Any theory that the increase represents "just cancers from AIDS" is false, because of the sharp rise from 1979 to 1983, before the AIDS outbreak began. Likewise, the change is not "just better detection of breast cancer cases" because of the rises among males. The fact that the 35 to 44 age group changed from no Boomers to all Boomers between 1979 and 1988 stands out, and points the finger at immune inadequacies of those born after World War II.

The National Cancer Institute study of cancer near nuclear plants also provides incidence data for Baby Boomers reaching their middle years. Such persons suffered a double-dip of radiation exposure, from bomb test fallout in childhood and emissions from nearby reactors as young adults. The NCI combined all persons age 20 to 39 as one category.

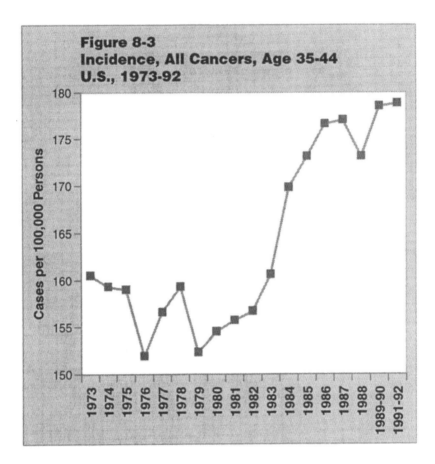

Figure 8-3
Incidence, All Cancers, Age 35-44
U.S., 1973-92

For both Connecticut and Iowa residents living in counties in which a nuclear reactor is located, unusual patterns of cancer incidence in adults 20 to 39 are emerging. The four nuclear plants in these two states opened between 1967 and 1974; thus, persons 20 to 39 in the years before the plant opening were largely born before the atomic era, while those born after the opening are mostly Baby Boomers.

Cancer rates in adults age 20 to 39 rose faster in the reactors' home counties than in the rest of the state. Cancer cases in New London County (CT) moved from 16% below to 8% below the state rate; Middlesex County (CT) went from 11% below to 3% above; and Benton/Linn counties (IA) jumped from 9% below to 7% above (see Table 8.1). Of the total of 1337 cancer cases in the three counties after plant startup, about 12%, or 160 cases are excessive for the 10 to 15 year period.

Breast cancer

Cancers known to be sensitive to radiation exposure again show the Baby Boomers in a bad light. Breast cancer in middle-aged women went through

Table 8.1 Changes in Cancer Incidence Rates Age 20–39,
Before and After Plant Startup

County	Cases		% Above/Below State Rate		% Change
	Before	After	Before	After	
All cancers except leukemia					
Middlesex CT	206	390	−11	+3	+14
New London CT	496	571	−16	−8	+8
Benton/Linn IA	128	376	−9	+7	+16
Female breast cancer					
Middlesex CT	48	71	−2	−1	+1
New London CT	90	101	−16	−7	+9
Benton/Linn IA	16	76	−4	+24	+28
Thyroid cancer					
Middlesex CT	7	23	−34	−14	+20
New London CT	16	41	−46	−6	+40
Benton/Linn IA	9	29	−11	+6	+17
Bone and joint cancer					
Middlesex CT	1	4	−57	+1	+58
New London CT	9	10	+29	+42	+13
Benton/Linn IA	0	4	—	+21	—
Leukemia					
Middlesex CT	8	16	−22	−7	+15
New London CT	19	26	−32	−10	+22
Benton/Linn IA	2	19	−60	+17	+77

Before,after startup = 1950 to 67, 1968 to 84 (Middlesex); 1950 to 70, 1971 to 84 (New London);
1969 to 74, 1975 to 84 (Benton/Linn)

the roof in the 1980s; some see this rise largely as the function of better diagnostic methods, while others contend that a true increase is actually happening. For women in their 40s, experts are split over whether the test is needed, but the percent of women in their 40s with a mammogram in the past 24 months rose from 31% to 61% between 1987 and 1994.[1] However, few women under 40 have mammographies, and any change in breast cancer rates under 40 closely represents a true change, not better detection. For Connecticut women age 25 to 39 born in the 1950s, 1245 were diagnosed with breast cancer, making the breast cancer incidence rate *17.9% higher* than among those born in the 1930s.

Unusually high increases in breast cancer for young women age 20 to 39 recently occurred near nuclear plants. After the Haddam Neck plant opened in 1967, the rate in Middlesex County barely moved, from 2% below to 1% below the Connecticut rate. However, New London County's incidence climbed from 23% below to 7% below the state average. In Iowa, rates for Benton and Linn counties rose from 4% under to 24% over the state standard.

Thyroid cancer

Thyroid cancer is also sensitive to radiation, but incidence rates are heavily influenced by children who underwent therapeutic head and neck irradiation, usually in infancy, between the late 1920s and late 1950s, when the documented risk of thyroid cancer caused physicians to cease the practice. With each succeeding decade, rates moved higher and higher. Even though a drop in thyroid cancer among persons born after the mid-1950s was expected, this did not happen. In Connecticut, the incidence rate for persons 20 to 39 born in the 1950s (based on 436 cases) was *64% greater* than for those born in the 1930s (2.55 versus 4.33 cases per 100,000 population). Thus, the prediction of less thyroid cancer for those born after 1950 has not materialized, meaning factors other than head and neck irradiation are at work to keep rates as high as they are.

Thyroid cancer increases among middle-aged Boomers are documented far beyond Connecticut's borders. In five states (Connecticut, Iowa, New Mexico, New York, and the western part of Washington), the incidence rate moved up 8.4% in the late 1980s, and 19.3% in the early 1990s. This trend mimicked the change in childhood thyroid cancer in Belarus and the Ukraine after the 1986 Chernobyl disaster, on a much smaller scale.

In the Connecticut and Iowa counties in which nuclear reactors are located, increases in thyroid cancer among those age 20 to 39 were especially steep after the plants opened. Middlesex County's rate went from 34% below to 14% below the state average, New London rocketed from 46% *below to just 4% below,* and Benton/Linn moved from 11% below to 6% above the Iowa average. These are significant changes, implicating not just Boomers' exposure early in life to atmospheric bomb test fallout, but to power plant emissions as well.

Bone and joint cancer

Bone and joint cancers are known to be affected by radioactivity, especially bone-seeking chemicals like strontium and barium. Even though these types of tumors are quite rare, the age 20 to 39 incidence in the four home counties of the three power plants in Connecticut and Iowa increased sharply, even though only 18 cases are involved.

Leukemia

Leukemia incidence in young adults is much greater than in teenagers, making meaningful analysis possible. In each of the three areas housing a power plant, leukemia rates jumped after plant startup. Bone-seeking radionuclides are known to increase the risk of developing leukemia; and the double-jeopardy combination of those being exposed to bomb test fallout as youngsters plus power plant emissions as young adults appears to support such a connection.

Cancer mortality

As stated before, cancer deaths may be less revealing than cancer cases in assessing radiation exposure's effects. The rapid advances in cancer detection and treatment in the 1980s and 1990s, when the Boomers moved into middle age, may tend to mask the true effects of an environmental pollutant on death rates. However, it is still important to look at any trends in cancer mortality, both in the U.S. as a whole and in populations near nuclear plants.

All cancers

Although cancer death rates for young American adults continued to fall in the 1970s and early 1980s, a slowdown in the rate of decline began in 1983. Cancer death rate decreases from 1983 to 1992 were only about half of what they were from 1970 to 1983. Again, those caught in this slowdown were the group born in the two decades after the atomic era started. Tucked into this period is 1986, the year of the Chernobyl accident. U.S. cancer death rates for that year for persons age 25 to 44 were completely unchanged from 1985, giving rise to speculation that Chernobyl fallout may have hastened the deaths of vulnerable individuals already suffering from cancer.

Another Boomer group vulnerable to excess cancer deaths are those living near nuclear plants. Looking at people living in the individual county in which nuclear reactors are located is one way to approach the issue, as the NCI did in its 1990 study and as we have done in previous chapters. However, this may be a narrow approach that might understate the true effects of radioactivity. When reactors release radioactive chemicals into the atmosphere, there is no imaginary wall at the county line stopping these particles from being carried by the wind into other, more distant areas, entering the food chain via rain or snow, and being consumed by people in food and drink. Some radioactive chemicals with a short half-life disintegrate rapidly and may not travel very far; but long-lived elements can be transported many miles. This was demonstrated in the atmospheric bomb tests, whose fallout covered the entire country, including the east coast, over 2000 miles away.

At some nuclear facilities, measurements away from the nuclear plant prove that radioactivity moves great distances. In Oak Ridge, Tennessee, for example, which began producing fuel for nuclear weapons in 1943, seven monitoring stations 25 to 75 miles from the plant took measurements of the grass, soil, water, and sediment in 1965 and 1978. Technicians found that some radioactivity levels (of long-lived chemicals) in these distant areas were between 50% and 100% of what they were at the reactor itself.[2] Furthermore, it is agreed that greater proportions of radioactivity are directed in the "downwind" direction by prevailing winds, leaving the population living downwind at greater risk of radiation's harmful effects.

The two concepts that downwind areas are at greater risk and that harmful effects can occur in areas far beyond one county surfaced in a 1994 study of cancer mortality among whites in the Oak Ridge area by Joseph Mangano.[3] While the 1990 NCI study only looked at Anderson and Roane counties in Tennessee, where the plant is located, Mangano included all 94 counties situated within 100 miles of the Oak Ridge facility. Cancer mortality in the 100-mile radius increased well beyond the U.S. rate from the early 1950s to the late 1980s (30.6% vs. 5.1%), suggesting that effects of Oak Ridge emissions extended well beyond the home county's border.

In addition, Mangano divided counties into downwind (to the northeast of the plant) and upwind areas (to the southwest). The long and arduous analysis produced disturbing results, especially for middle-aged people. No difference existed between downwind and upwind trends for persons age 20 to 39, although both were in excess of the national rate. However, for those 40 to 44, 45 to 49, and 50 to 54 in the late 1980s, who were fetuses, infants, and young children during the early years of Oak Ridge's massive releases (1943 to 1948), a gaping hole between downwinders and upwinders exists, specifically:

- Age 40–44, +24% downwind, −16% upwind, −27% U.S.
- Age 45–49, +40% downwind, −3% upwind, −16% U.S.
- Age 50–54, +115% downwind, +13% upwind, −2% U.S.

The excess for 1987 to 1989 alone in the 33 downwind counties is about 200 of the 600 cancer deaths. Downwind cancer rates, which were below U.S. figures in the early 1950s, were about 45% higher in 1987 to 1989. The greater concentrations of radioactivity from Oak Ridge in downwind areas is likely to be driving up cancer rates. Upwind and downwind areas have the same cancer risks in terms of diet, poverty, access to medical care, air and water pollution, and even bomb test fallout.

Another, broader analysis of the cancer experience of middle-aged Baby Boomers and their predecessors uses the 1990 NCI study once again. A number of nuclear power plants started up operations in the late 1960s and early 1970s, just as the population age 20 to 39 began to be infiltrated by Boomers. Comparing cancer death rates before and after plant startup constitutes a rough comparison between Boomers and pre-Boomers.

Local death rates for adults age 20 to 39 for all cancers combined in nuclear counties didn't change very much for plants opened before 1970 and from 1970 to 1974 (Table 8.2). However, a different story emerges for cancers sensitive to radiation. Thyroid cancer deaths are few in number — this is perhaps the most curable of all cancers — but death rates for nuclear counties compared to the national rate jumped from 14% below to 16% above for plants open before 1970, and from 17% above to 57% above for those starting operations in 1970 to 1974. Exposure to strontium and other bone-seeking radioisotopes increases risk of bone and joint cancers, and rates rose after plant startup. Leukemia showed a significant rise for plants opened in 1970 to 1974, and a slight decrease for the earlier generation of power plants. Female breast cancer deaths showed no excess in counties near reactor power plants.

Table 8.2 Cancer Mortality, Age 20–39, Counties with Nuclear Power Plants, Before and After Plant Startup

Type of Cancer	Deaths		% Above/Below U.S. Rate		% Change
	Before	After	Before	After	
Facilities opening before 1970					
All, excl. leukemia	3225	5425	−6	−4	+2
Female breast	494	820	+ 6	− 1	− 7
Thyroid	10	18	−14	+16	+30
Bone + joint	56	118	−13	+3	+16
Leukemia	400	736	+4	+2	−2
Facilities opening 1970–74					
All, excl. leukemia	6104	3248	−1	−2	−1
Female breast	830	500	− 3	− 3	+ 0
Thyroid	24	13	+17	+57	+40
Bone + joint	126	80	+12	+18	+6
Leukemia	665	474	−4	+9	+13

Breast cancer

One type of cancer that may reveal more about radiation's effects on Baby Boomer mortality is breast cancer. The rapid rise in newly diagnosed breast cancer cases in the 1980s caused considerable fear among the population. The response of many in the scientific community was a cautious one, based on the "wait and see" principle to determine whether this rise is a one-time phenomenon, perhaps because of better means of detection through more frequent mammograms. The National Cancer Institute has touted slight reductions in breast cancer death rates for the entire female population in the early 1990s. However, women 25 to 44 benefit only minimally from better detection, since very few women under 45 undergo mammograms.

Trends in breast cancer deaths for women 25 to 34 are not very meaningful, because there are only about 600 annual deaths in the U.S. But for women 35 to 44, a population that loses 3500 of its members each year to the disease, three distinct trends unfolded during the atomic age (Figure 8.4). From 1951 to 1970, the rate was virtually unchanged; perhaps this occurred because of lack of therapeutic advances. From 1970 to 1978, rates fell steadily each year, dropping 17% in the 8-year period. Finally, as the first Baby Boomers entered the 35 to 44 group in 1979, the rate stagnated, and by 1991 still hadn't gotten back to the historic low set in 1978. Excess breast cancer deaths in the age group 35 to 44 from 1979 to 1991, based on the 1970 to 1978 drop as the expected trend, is about *5,000 to 10,000*. This puzzling scenario remains unexplained, and is added to the increasingly poor health record of the Baby Boomers.

The stagnation in breast cancer mortality rates has dropped the U.S., a country second to none in curative technologies, behind many other nations

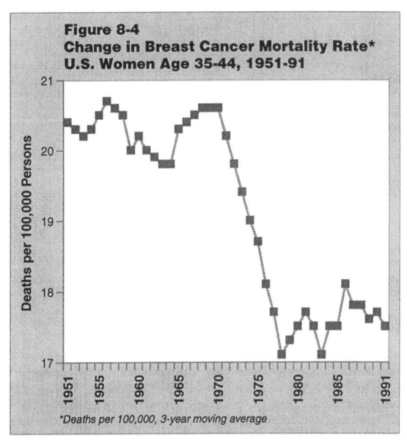

**Figure 8-4
Change in Breast Cancer Mortality Rate*
U.S. Women Age 35-44, 1951-91**

*Deaths per 100,000, 3-year moving average

of the world. The 1990 rate of 17.8 deaths per 100,000 among women 35 to 44 places the U.S. well behind the industrialized Pacific countries of Japan (9.2) and Australia (15.9). In addition, the U.S. trailed 8 of 11 Communist-bloc nations in eastern Europe, 12 of 18 western European countries, and 4 of 5 countries from Central and South America. These figures, compiled by the World Health Organization, further illuminate an unexpectedly high breast cancer death rate among Boomer women, a fact that cannot be explained by industrialization alone, because many countries faring better than the U.S. are industrialized and polluted as well.

Chronic fatigue syndrome

As the Boomers were assaulted by the AIDS epidemic in the latter years of the 20th century, another new immune scourge also reared its ugly head. In the winter of 1984–85, reports began to filter out of Incline Village, Nevada, a small town near Lake Tahoe, of a large number of people feeling very fatigued and experiencing a variety of other symptoms for long periods of time. Eventually, about 200 in this small town came down with the illness.

Very soon after, reports of similar cases began to surface, not in a cluster like Nevada's but spread out over the entire nation and overseas. Doctors found that some patients had antibodies to the Epstein–Barr Virus, and the press was quick to jump on the lead, calling the disease Chronic Epstein–Barr Virus syndrome. However, the large numbers of sick people with the disease who tested normal for EBV soon dispelled this title; and for lack of a better name, the condition was dubbed chronic fatigue syndrome (CFS).

The disease proved to be a major drain on many who developed it. Far beyond simple fatigue, which many Americans accept as a normal part of life and which can be erased by reduced stress, more rest, and more exercise, CFS proved to be much more. Prolonged rest often did little or nothing to bring patients back to health, and exercise often made matters worse. Furthermore, a variety of symptoms contributed to patient miseries. Many experienced cognitive problems such as inability to focus, concentrate, remember, or sleep properly; neuromuscular ailments such as painful or achy joints and muscles; and assorted other symptoms such as nausea, fever, tender lymph nodes, and frequent sore throats.

In the late 1980s, the exploration of CFS was dogged by a skepticism among the American public, fueled by some cynical physicians and largely apathetic public health officials, that the disease was a hoax. Some physicians quickly dismissed the illness as just depression, or a psychosomatic disorder in which patients "wanted" to feel sick. The press added to the controversy by repeating the appellations "yuppie flu" or "affluenza." But those physicians who took the time to thoroughly examine and test patients knew that the disease was a lot more than unhappiness. Depression's epidemiology is different than CFS (i.e., it never occurs in clusters); the symptoms aren't the same; lab abnormalities are different; and depressed patients respond differently to treatment than those with CFS.[4] Many patients were found to have a variety of immune abnormalities; prominent among these is a low number and low toxicity of Natural Killer (NK) cells, a type of white blood cell crucial to fighting cancer and infections. The disease is now known by many knowledgeable patients and physicians as Chronic Fatigue and Immune Dysfunction Syndrome, or CFIDS.

Despite working with a severe paucity of research dollars, scientists have uncovered immune troubles in CFIDS that are a combination of immune suppression (like in AIDS), and immune up-regulation, or an overactive immune system. Luckily, the up-regulation component appears to be the stronger of the two, and the suppression has not been predominant enough to cause deaths or allow opportunistic infections common in AIDS patients. Although there may be some connection to AIDS (some believe reasons causing them might be similar), these are two wholly separate diseases.

The immune up-regulation findings include sensitivity to allergies in many CFIDS patients; production of large numbers of autoantibodies against an invading virus or infection; increased numbers of T8 lymphocytes, a component of the immune response; and activation of chemicals such as interleukin-2 and interferon, also reflecting the "switched on" immune system looking for an outside invader to attack. Examples of immune suppression include activation of certain viruses such as Epstein–Barr or Human Herpesvirus-6 in some patients; presence of the common yeast infection *Candida*; low levels of immunoglobulins, part of the fighting immune system; little or no response (cell-mediated immunity) to skin tests of certain antigens; and as mentioned, low numbers and low functional levels of Natural Killer (NK) cells.[5]

All patients have a different profile, which adds to physician frustration because of the elusiveness of pinpointing just what the disease is and what can be done about it. However, there is a particular concern about the last finding, low NK cells, because this is a key aspect of the body's ability to fight off cancer cells and herpes viruses. Of all the abnormal findings, low NK cell count and function seems to be the most consistent among the CFIDS population (in Japan, the disease is known as Low NK Cell Disease). Luckily, after over a decade of experience with CFIDS, no unusual numbers of cancers have appeared among persons afflicted with the disorder; but many physicians and patients are holding their breath that no cancer outbreak hits CFIDS patients in the future.

Boomers were squarely in the middle of the CFIDS epidemic from the start. The earliest research of the U.S. Centers for Disease Control in the late 1980s confirmed that about 75% of the cases were in persons age 20 to 44 (born 1945 to 1970), roughly similar to AIDS.[6] Determining the number of CFIDS cases in the U.S. has been a difficult task, hampered by a lack of a clear definition, apathy of physicians and public health professionals, and a gross shortage of research dollars. However, by the early 1990s, estimates began to emerge, and they were startling. One study of a large health maintenance organization found the prevalence to be between 75 and 267 per 100,000.[7] The Centers for Disease Control and Prevention, which had been accused of underestimating the body count, agreed, estimating the number between 76 and 233 per 100,000.[8] If this estimate is accurate, roughly 150,000 to 500,000 American adults, plus a small number of children, have the disease. Another study estimates that up to 390,000 U.S. adults have CFIDS.[9] Thus, approximately *100,000 to 350,000* Baby Boomers suffer from the disease, about the same number living with full-blown AIDS.

Is CFIDS a new disease or the continuation of an old one? The answer is still uncertain, but many experts seem to agree that it existed prior to the 1980s, documenting numerous outbreaks of a strange, flu-like fatiguing illness, most since 1948.[5] Virtually no data on trends in CFIDS have been accumulated thus far. However, one large CFIDS practice in Charlotte, North Carolina, known throughout the U.S. suggests that there was a rise in the 1980s. Using data from over 1200 CFIDS patients who had consulted the practice, an explosion of new cases occurred from 1980 to 1987, with the annual number of cases roughly leveling off after 1987 (Figure 8.5). The study was made at the end of 1994, but because patients often have the disease for several years or more before getting to Charlotte, allowance must be made

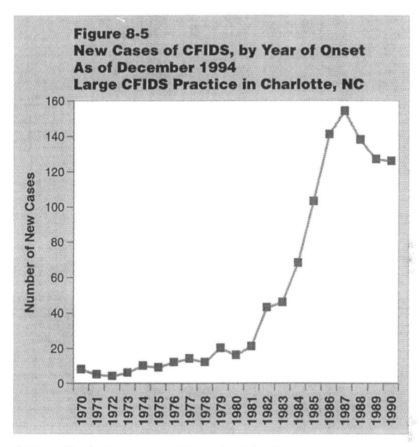

Figure 8-5
New Cases of CFIDS, by Year of Onset
As of December 1994
Large CFIDS Practice in Charlotte, NC

for this lag. Charles Lapp, one of the nation's leading CFIDS experts believes "there was definitely an increase from 1980 to 1987."[10] He points out that the curve is similar to new AIDS and breast cancer cases in the 1980s, although he isn't sure why.

Another suggestion that CFIDS cases are rising comes from an article in the business section of the November 28, 1994, *New York Times*. The article cites "Epstein–Barr Virus" as the disease accounting for the second-fastest rise in disability claims to UNUM, the largest private group disability insurer in the U.S. The number of such claims grew from 40 to 243 from 1989 to 1993, according to the company.[11] Cruelly, the article quoted a company executive complaining about a growing number of "nervous- and mental-type situations," and cited a cynical insurance analyst in a large Wall Street firm as accusing white-collar professionals, especially doctors, of "using disability insurance as a substitute for unemployment compensation."

Because of the lack of research funds, any potential link between Baby Boomers, radiation exposure, and CFIDS remains untouched. However, the issue is a live one for several reasons. First, the onslaught of other immune diseases in the Baby Boomer generation, who were the first to come of age in the nuclear era, merits scrutiny of all potential environmental insults to the immune system, particularly new ones such as radiation. Second, a peak of new cases may have occurred around 1986 to 1987, in the aftermath of Chernobyl. Third, there is evidence that some immune abnormalities in CFIDS patients are those with proven links to radiation exposure. Perhaps the most telling of these links involves NK cells; in a 1977 article, Swedish researchers fed strontium-89 to rats, and measured a corresponding decrease in NK cell number and potency.[12] strontium-89 is a fission product consumed in substantial amounts during the years of atmospheric bomb testing.

Lyme disease

In the late 1980s, another new disease caught the eye of the public and health professionals. Reports mostly from the northeast U.S. described a condition called Lyme disease, named for the Connecticut town with an unusually high incidence. Unlike CFIDS, the cause of Lyme disease was quickly uncovered. Ticks, which had picked up a bacterium called *Borrelia burgdorferi* from infected deer, bit and infected unsuspecting individuals. The disease that often ensued usually consists of a rash in the shape of a bulls-eye, malaise, fever, neurological problems, joint pain, headache, and stiff neck. Fortunately, antibiotic treatments often prove effective if the disease is spotted in time.

In the early 1990s, just over 10,000 new cases per year were reported in the U.S., about 75% in New York, New Jersey, Connecticut, and Pennsylvania. Roughly 40% of the cases were persons in their 20s, 30s, and 40s, again pulling Boomers into the teeth of a new immune-related epidemic. Not enough research has been done, but impaired immune systems that have reduced ability to fight off invading bacteria may implicate radiation exposure as a possible cofactor in the spread of Lyme disease. Such a scenario can be traced back to Andrei Sakharov's original thesis in the 1950s that nuclear fission products would raise human sensitivity to bacterial infections.

Allergies

Allergies, or human sensitivity to natural substances, are certainly not a new phenomenon, nor do they exist only in America. However, the level of a person's sensitivity depends on how strong the person's immune system is to fight off the offending substances, such as pollen.

Allergy prevalence in the U.S. consists merely of estimates, and trends are even more of a guess. However, many experts agree that the number of allergy sufferers in the U.S. has increased since the 1970s. The only surveys on prevalence by age group that exist are sporadic, one-time efforts, making trend analysis impossible. But even these studies are revealing. The National Center for Health Statistics performed skin tests for eight common allergies on a sample of 15,000 Americans in the late 1970s. For the three most common (house dust, rye grass, and ragweed) persons between 12 and 34 (almost all Boomers) had the highest rates of positive reactions (Figure 8.6). Among Boomers, those born in the years between 1954 and 1960, when the most

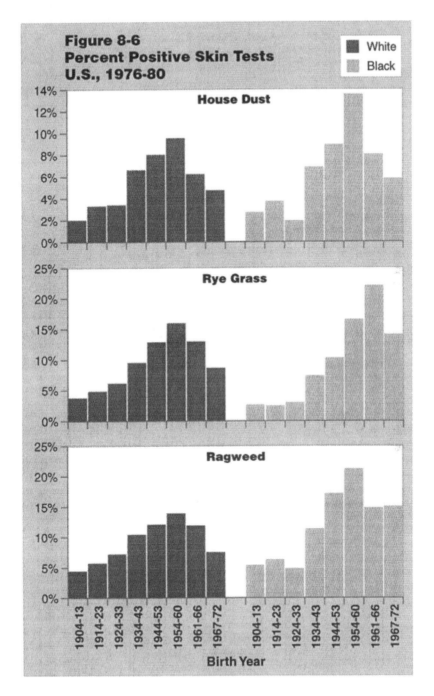

Figure 8-6
Percent Positive Skin Tests
U.S., 1976-80

House Dust

Rye Grass

Ragweed

Birth Year

powerful atomic bombs were tested above the Nevada landscape, often had the highest rates. Post-Boomers age 6 to 11 in the late 1970s (born between 1967 and 1972) had consistently lower rates than did the Boomers.

ABC's Sam Donaldson reports experts believe that one reason for the spread of allergies "is that we're healthier than ever; when our immune systems aren't busy fighting disease, they're more likely to fight substances to which they're allergic."[13] This statement collapses amidst the rash of rising immune diseases that depend on robust immune responses boring in on the American population, especially the post-war generation.

Asthma

Asthma is a disease of the respiratory system, marked by inflammation or obstruction of the airways. It is a common disease affecting an estimated 9 to 12 million Americans, and is most prevalent among children. Asthma is a type of allergy; thus, the failure of the body to handle allergens such as ragweed, pollen, and dust makes a person more susceptible to asthma.

Advances in medical treatments over the years have made asthma a generally controllable disease, even though it still accounts for many days of school or work missed, along with considerable discomfort to victims as they struggle to breathe. The Asthma and Allergy Foundation of America estimates that the rate of Americans with the disease rose about one third during the 1970s and the 1980s, and hospitalizations for asthma rose 10% from 1987 to 1991. Unfortunately, these are just estimates, and how the Boomers' incidence fared compared to other age groups is not precisely quantifiable.

Although deaths due to asthma are rare, they sometimes occur in serious or untreated cases. Once again, the meticulous collection of yearly data on mortality since the 1930s enables analysis of how the disease is affecting Baby Boomers as they move into adulthood. The results are startling. In the 1960s and 1970s, asthma rates for persons age 25 to 44, representing babies born during or before World War II, dropped by about 70%, reaching a low in 1977–78. However, since that time, the trend has reversed, and the death rate has climbed sharply. From 1979 to 1981 (when the 25 to 44 age group consisted of persons born from 1936 to 1955) to 1989 to 1991 (born between 1946 and 1965, all Boomers), the asthma death rate rose 63.2% (Table 8.3). While fewer than 300 young adults died of the disease each year in the late 1970s, now about 700 such deaths occur, and the increase shows no signs of slowing down. Asthma deaths are rising for all age groups, but none faster

Table 8.3   % Change in Death Rates for
Four Immune-Related Diseases,
U.S., Age 25–44

Birth Year	Asthma	Congenital Anomalies	Septicemia	Viral Hepatitis
1926–45	−38.7%	−18.2%	+36.2%	+1.1%
1946–65	+63.2%	−11.8%	+81.3%	+71.7%

than the Boomers. Again, asthma risk is higher in allergic persons, whose immune systems are unable to cope with the offending substances. This explosive epidemic is not well known to the public, but it threatens society, and especially the Baby Boomers, with more disability, more hospitalizations, and sometimes more deaths. In addition, it may provide another clue that the immune systems of the post-war generation may be functioning below par.

Congenital anomalies

As discussed in Chapter 6, congenital anomalies, or birth defects, can cause death in all ages, although most victims are infants. Among young adults, dying from a residual effect of a birth abnormality is not a common happening, with fewer than 1000 victims age 25 to 44 per year in the U.S. Still, historical data provide some interesting trends. In the years following World War II, congenital anomaly death rates for persons age 25 to 44 were on the rise; interestingly, from 1951 to 1963, the years of atmospheric bomb testing, the rate rose 41%, from 1.514 to 2.135 per 100,000 persons, even though people in this age group had been born long before the atomic era, and suffered no damage from fallout during infancy and childhood.

After 1963, death rates fell consistently, and by 1980, the rate was about half what it was in 1963. However, a reversal in the death rate began in the mid-1980s; by 1989–91, when the 25 to 44 age group represented the babies born from 1946 to 1965, the drop was slowing, and actually rose after 1984. This reversal is not as pronounced as that of asthma deaths, but the known link of radiation exposure to congenital anomalies makes this finding one of concern.

Septicemia

Septicemia, or blood poisoning from bacterial invaders, was also discussed in Chapter 6. Although the disease is most likely to strike the elderly, it affects all age groups, including middle-aged adults.

The U.S. is in the midst of a prolonged epidemic of septicemia deaths, especially for young adults. In the time immediately following World War II, the death rate was dropping, aided by the development of new antibiotic drugs to fight off infections. However, the death rate has been on the rise for young adults since the 1950s, and the rise accelerated during the 1980s. From 1979–81 to 1989–91, the U.S. septicemia death rate for those age 25 to 44 rose 81.3%; the annual number of deaths in each of these two periods zoomed from 328 to 894. A rise in septicemia mortality among elderly persons has a possible explanation in that the immune response among older persons is often decreased. However, the rapid and (thus far) unchecked rise among young adults remains unexplained.

Viral hepatitis

Another disease that has shown a troubling rise in the 1980s and 1990s among young adults is viral hepatitis. This disease causes the liver to become inflamed and dysfunctional. It often requires hospitalization and may cause death in some cases.

The mortality rate for viral hepatitis among young adults, after years of little progress, dropped sharply in the 1970s. By 1979–81, the rate of 0.2567 per 100,000 was well below the 1969–71 mark. However, the progress abruptly stopped about 1978, and rates began to rise. By 1989–91, the rate had soared from 1979–81, and the number of young Americans succumbing each year had surpassed 400, up from about 150 in the late 1970s.

Obesity

The issue of body weight is ubiquitous in modern American society. A virtual industry has grown up around fitness clubs, athletic attire, diet programs, and low-fat, low-calorie foods. A virtual obsession with weight gripped the nation in the late 20th century.

Despite this intense interest, Americans are doing worse when they step on the scale as the years go on. In 1994, information reported in the *Journal of the American Medical Association* showed an alarming increase in the percentage of overweight adults from the early 1960s to the early 1990s. For Baby Boomer women, the proportion of those considered overweight rose 100% for those age 20 to 29, 57% for those 30 to 39, and 40% for those 40 to 49.

Among pre-Boomers, the rate for women 50 to 59 rose 49% but *decreased* 9% for women 60 to 74. The article acknowledges that most of the rise in overweight Americans occurred in the 1980s, after minimal change in the 1960s and 1970s. Rates for Boomer men also increased, but at a slower pace.[1]

At first glance, being overweight would seem to have nothing to do with one's history of radiation exposure. However, radioactive iodine, one of the primary components of bomb test fallout and nuclear plant emissions, is known to promote weight gain by impairing the thyroid gland (which regulates metabolism) and making the affected person lethargic. Even small doses can cause this effect: radiation expert Rosalie Bertell says that "a mild exposure experienced by a large population could cause a decrease in average thyroid hormone levels and an increase in average body weight, such as is occurring now in the North American population."[14] Furthermore, female Boomers experienced greater weight gain than men; coupled with the fact that the majority of radiation-linked thyroid problems (cancer, hypothyroidism) occur in women, a troubling shadow is cast over body weight and overall metabolism by radioactive iodine exposure, which affects these two clinical conditions.

Depression

Another health problem that worsened in the 1980s and 1990s is depression. Mental health conditions have historically proved difficult for doctors to pinpoint accurately; diagnosing depression is not nearly as easy as detecting diabetes or a broken leg. However, the criteria for depression are generally agreed upon; hopelessness and helplessness are the primary markers, along with inability to concentrate or remember things, appetite changes, sleep disorders, and loss of interest in life's activities. Severe cases can be accompanied by thoughts of suicide. Scientists have recently made substantial gains in expressing the physiological abnormalities of depression, and have developed medications to counteract them. For example, many depressed persons are known to have low levels of the chemical serotonin in the brain. A family of medications known as selective serotonin reuptake inhibitors that increase serotonin levels have become extremely popular and effective; drugs such as Prozac, Zoloft, and Paxil have almost become household names in America.

More evidence is being uncovered that physical as well as emotional factors affect the disease. Postpartum depression suffered by a number of new mothers is typically a function of bodily changes. The onset of winter, with less sunlight, may affect certain chemical processes, leading to depression. Certain diseases or medications carry with them an increased risk of depression.

A number of articles in the lay press have picked up on a perceived trend that depression is on the rise among young adults. "Various researchers have reported that the rate of depression among Baby Boomers is rising," states a recent article in *Esquire* magazine.[15]

Scientific studies also show Boomers to have more depression than would be expected. In 1989, two New York researchers culled together the results of separate studies from five areas (Baltimore, MD, Durham, NC, Los Angeles, CA, New Haven, CT, and St. Louis, MO). Trends in each of the five sites "found an increase in lifetime rates of major depression in the younger age cohort, usually those born after 1940," write the researchers. At age 24, about 0.5% of persons born from 1905 to 1934 had developed depression, compared to 3% for those born from 1935 to 1944, 5% for births between 1945 and 1954, and 7% for those born from 1955 to 1964 (Figure 8.7). Thus, those born from 1955 to 1964 (the height of atmospheric bomb testing) had a *14 times greater* chance of becoming depressed by age 24 than most pre-World War II births. This gap is consistent for those age 34 as well.[16] The authors agree that six factors can cause depression, including environmental pollutants and genetic factors, leaving room for speculation that radiation might be one of the environmental causes of this trend.

The ability of radiation exposure to compromise human immune cells and alter hormones in the body may possibly be one type of change that raises a person's susceptibility to depression.

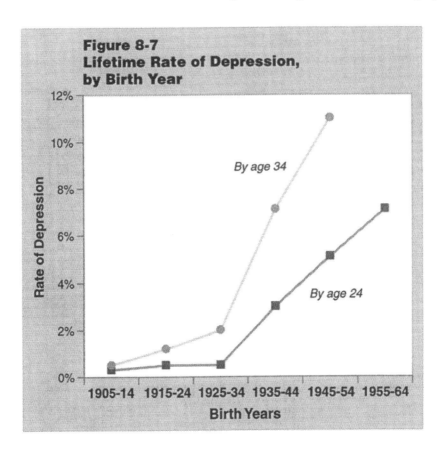

**Figure 8-7
Lifetime Rate of Depression,
by Birth Year**

Childbearing

Underweight births

As the Boomers grew up and became sexually active, they began to produce a new generation of Americans. The health status of the post-Boomer generation will be examined in the next chapter. A parent's health status is often an important clue as to how healthy the child will be, although establishing exact links is difficult. When a child develops a disease like cancer or pneumonia, susceptibility from a parent's genes may play a role in the disease.

As we shall see in the next chapter, low-weight births reversed their downward trend beginning in the mid-1980s and began to rise. By 1992, about 60% of newborns had Baby Boomer mothers, who were between 25 and 44. Between 1984 and 1992, rates of underweight births for whites with Boomer mothers rose slightly, while rates for blacks with Boomer mothers went up rapidly (Table 8.4). Underweight birth rates dropped slightly for both white and black babies who in 1992 had mothers born after the Baby

Table 8.4 % Change in Rates
of Births Under 2500 g,
U.S., 1984–92

1992 Birth Year	White	Black
1973–77	–2.5	–2.9
1968–72	–0.0	–0.0
1963–67	+2.0	+10.1
1958–62	+5.8	+25.6
1953–57	+10.2	+26.6

Boom. With over 4 million births in America each year, the numbers are so large that this contrast could not have happened by chance. The problem of more underweight births in the late 1980s and early 1990s lies strictly with the children of Baby Boomers.

Assisted pregnancy

As seen in the previous chapter, infertility rates in Baby Boomer women were on the rise as early as the 1970s. In the following decades, however, clinicians devised a number of methods to assist couples experiencing difficulty in their attempts to conceive a child. One of these approaches is *in vitro* fertilization, in which the man's sperm and the woman's egg are fertilized outside the body before injecting the combination into the woman. *In vitro* fertilization fails far more frequently than it succeeds and is quite costly, but has become increasingly popular. Perhaps the most comprehensive study on the procedure's success was released in December 1997 in the form of a Centers for Disease Control and Prevention report on 281 U.S. clinics' 1995 performance.[17]

While the report compared success rates for each clinic, it also looked at national statistics, including pregnancy rates according to the age of the mother. If only those "cycles" (attempts to conceive) using the mother's eggs and not a donor's are considered, the percentage of live births per cycle for mothers under 35 (25.3) is larger than that for mothers 35 to 39 (18.2) and mothers over 39 (8.0). A total of 45,000 cycles were included in this comparison. Naturally, increased maternal age makes it harder to achieve a successful pregnancy, but maybe this does not explain the entire reason for the large gap. Once again, birth year may play a very important role in understanding differences. Women under 35 were born during or after 1961, making most of them post-Baby Boomers. Women over 39 (most are probably between 40 and 45) were born from 1950 to 1955, the Boomers who lived through the entire Nevada testing period. No exact calculation can be made, but the fact that younger women had three times the success rate as their older counterparts (25% to 8%) raises questions about the women's ability to conceive aside from their age.

Sperm concentration

Another measure of the health of the parent that has not traditionally been of concern to public health and medical professionals is fertility. The inability of some couples to bear children is an age-old problem, but most men and women have been fertile enough to continue populating the world and extending life. Until recently, medical science could do very little about the problem. In recent years, advances to assist infertile men and women have introduced almost miraculous techniques such as *in vitro* fertilization, fertility drugs, and sperm banks.

Because this new "industry" of technology to counter infertility developed, a new means of assessing infertility emerged. In the late 1980s, Danish physician Niels Skakkebaek found that it was more and more difficult to find men who could donate acceptable semen to sperm banks for the infertile. This puzzling development led Skakkebaek to begin looking at sperm counts, defined as the concentration of sperm in the semen. In 1992, Skakkebaek published a review of 61 studies from all over the world and found that the rate in Denmark had declined an astounding 42% from 1944 to 1990, from 113 million to 66 million sperm per milliliter of semen. Studies of U.S. males reported similar declines.[18] A team of scientists in Paris, unable to believe this remarkable finding, set out to do their own research. They obtained the same results in 1995; 30-year-old French men in 1992 (born 1962, at the height of atomic bomb testing) had an average sperm count of 51 million, *only half* that of 30-year-olds in 1975 (born 1945, just as the atomic age was about to begin).[19] In addition, the percent of men with fertility problems due to low sperm concentration has tripled since the 1940s, from about 6% to 18%. Any sperm count under 20 million is considered "subfertile," while counts under 5 million usually mean sterility. The studies also show that in addition to lower sperm counts, the quality of sperm has suffered a decline in the latter 20th century.

These discoveries were challenged by some scientists, as is customary in any new and dramatic finding. However, other scientists supported the conclusions and began speculating on the causes; environmental factors immediately took the spotlight. Suddenly, stories of increasing infertility began to crop up in the news. The National Center for Health Statistics says that the percentage of infertile couples has grown from 14.4 to 18.5 between 1965 and 1995.[20] A more careful look at this trend places Baby Boomers squarely at the epicenter of the problem once again. In the previous chapter, we have seen that infertility in women between 1965 and 1976 was rising only in women under 30, and falling in women over 30. In 1976, women 15 to 29 were Boomers, while those older than 30 were born earlier. This sharp contrast suggests that a *current* environmental problem is *not* the only cause, but something that affected only particular age groups. Radiation must be considered as a potential clue to solving this new and most troubling mystery. A radioactive chemical such as strontium-90 and its daughter product yttrium-90 are known to damage the hormonal system and to concentrate in the sperm and damage cells therein.[21]

A trend that parallels the decline in sperm counts is testicular cancer. This type of cancer is usually very treatable and rarely fatal, and is found mostly in men age 20 to 44. From 1973 to 1992, the incidence of testicular cancer in U.S. men rose 53%, from 3.0 to 4.6 per 100,000. By 1992, most victims of this type of cancer belonged to the Baby Boom generation. A rise of about 50% in testicular cancer and a decline of about 50% in sperm concentration is too stark a pair of trends to dismiss. Of all the health effects the Baby Boomers have suffered or are still suffering, this may be the most disturbing one, because a continuation of these patterns affects the ability of Americans to pass along good health or even life itself to future generations.

Disability

Another measure of health in a population is what proportion are disabled. Only estimates, rather than an exact count, exist on how many Americans are disabled. However, a large-scale disability database is maintained by the federal Social Security Administration. While Social Security still serves mostly aged Americans, it has offered benefits to disabled persons under 65 years old since 1957; and in 1972, disability recipients on the rolls 18 months or longer (most recipients) became eligible for Medicare benefits.

From 1980 to 1987, Medicare enrollment for persons under 65 (mostly disability recipients) remained at 3 million, due in part to the draconian measures instituted by the Reagan administration to cut expenses by restraining program eligibility. From 1987 to 1996, however, the rolls grew 53%, from 3.0 to 4.6 million Americans.[22] But perhaps more important, persons age 35 to 54 accounted for virtually all of the spiraling growth. Table 8.5 indicates that between 1984 and 1994, disabled Medicare recipients age 35 to 44 grew 104%, while those 45 to 54 advanced 77%. All others under age 65 only increased by 9%.[23] In the 10-year span from 1984 to 1994, the 35 to 44 age group moved from those born in the 1940s to those born in the 1950s, the years of the most hazardous atomic tests over the Nevada desert. Those 45 to 54 changed from those born in the 1930s to those born in the 1940s (at least half were born after the atomic age began).

Table 8.5   Change in Disabled Persons over 65 on Medicare, U.S., 1984–94

Age Group	Birth Year		No. on Medicare (in Thousands)		%Ch.
	1984	1994	1984	1994	
35–44	1940–9	1950–9	422	859	104%
45–54	1930–9	1940–9	584	1033	77%
All other	—	—	1878	2051	9%

Once again, Boomers are implicated in an adverse health trend. But even though we are not examining a specific disease like leukemia or asthma, the explosion of Baby Boomers on the disability rolls may be more disturbing, because it represents a sidelining of large numbers of society's contributors in their peak years. The costs are alarmingly high, in terms of young persons being robbed of their productivity and of the monetary loss to society. Combining the total of disability payments, Medicare expenses, and lost tax revenue, a total of *about $15 to $20 billion each year* is lost to society from the additional 886,000 disabled Americans 35 to 54 on Medicare. Statistics like this make it easier to understand why governments have run deficits in recent years, why productivity in American business has slumped, and why Social Security and Medicare enter the future in precarious financial positions.

Summary

After a sub-par performance in rates of immune-related disease as infants and children, the Boomers took a tremendous thrashing as young adults. In the 1980s and 1990s, perhaps 1 million Boomers were stricken by the relatively new diseases of AIDS and CFIDS, which either caused serious immune damage and/or overpowered susceptible immune systems. The overall death rate for persons age 25 to 44 broke a decades-long decline about 1983 and started to increase. Although this turnaround is largely due to AIDS, there is evidence that increases were especially rapid near several old nuclear plants.

Cancer incidence in young adults was higher for Boomers than for their parents. Some especially disturbing trends took place for women in their 30s and 40s; breast cancer incidence soared after 1980, and breast cancer mortality rates stagnated after 1978. Rates of radiation-sensitive cancers rose especially quickly in Boomers living near nuclear plants.

A series of negative reproductive trends were another mark against the post-war generation. Infertility rose, while Boomer mothers began bearing larger numbers of underweight babies. Young men were hit by dramatically lower sperm counts and higher rates of testicular cancer. Finally, the Social Security disability rolls began to bulge with disproportionate numbers of Baby Boomers, costing Americans billions of dollars each year. Again, these adverse trends are influenced by more factors than just radiation exposure, including better diagnostic methods, access to medical care, and personal lifestyles, but an increasing number of scientific experts are looking toward environmental causes to explain at least part of the adverse trends. It appears that radiation exposure is a not-insignificant part of this environmental component, especially in connection with other carcinogens working together in a synergistic relationship. For example, we know that uranium miners who smoke are at greater risk for certain cancers than those who don't smoke. Smokers living in houses with high radon levels have higher disease rates than non-smokers in the same houses. And again, there is no single alternative explanation for excess disease and deaths in persons living near nuclear plants, other than exposure to radioactive emissions from the plants.

The toll of Boomers from immune disease mounted, and the damage could be viewed as more than just a death toll. Medical bills to treat patients with cancer, AIDS, CFIDS, and infertility climbed into the billions of dollars. Instead of reaping the benefits of young adults, in the prime of their life, contributing to the betterment of society, society had to assume the expense of caring for them. Instead of vibrant contributors, these same persons were taken from society by death or turned into inactive humans taking from society in the form of health expenses and disability checks. Some of the best and brightest of a generation became disabled or died, and American business began to fall behind other nations, such as Germany and Japan, in terms of innovative ideas and commercial growth.

As the oldest Boomers reached 50 and glimpsed the years of much higher disease rates, the immune disease record for those born from 1945 to 1964 hovered like a black cloud over the generation's future.

References

1. National Center for Health Statistics, *Health United States 1996–97 and Injury Chartbook*, DHHS Publication (PHS) 97-1232, Hyattsville MD, 1997.
2. Oak Ridge Health Assessment Steering Panel, *Oak Ridge Health Studies Phase I Overview*, Vol. II, Part A, Nashville, TN, 1993.
3. Mangano, J., Cancer mortality near Oak Ridge, Tennessee, *International Journal of Health Services*, Summer 1994, 521-533.
4. Lapp, C., Chronic Fatigue Syndrome is a *real* disease, *North Carolina Family Physician*, Winter 1992, 6-11.
5. Bell, D., *The Disease of a Thousand Names*, Pollard Publications, Lyndonville, NY, 1991, 84, 93-97.
6. *Heart of America News*, Fall/Winter 1990, 2.
7. Buchwald, D., et al., Chronic fatigue and the Chronic Fatigue Syndrome: prevalence in a Pacific Northwest health care system, *Annals of Internal Medicine*, July 15, 1995, 81.
8. CDC announces new prevalence data, *CFIDS Chronicle*, Spring 1995, 34.
9. *CFIDS Chronicle*, Winter 1995, 49.
10. Personal correspondence, Charles Lapp, December 1994.
11. Personal correspondence, Tracy Sherman, UNUM Corporation, Portland, ME, November 29, 1994.
12. Haller, O. and Wigzell, H., Suppression of natural killer cell activity with radioactive strontium: effector cells are marrow dependent, *The Journal of Immunology*, 1977, 1503-6.
13. Sam Donaldson, on *Prime Time Live*, ABC Television, May 24, 1995.
14. Bertell, R., *No Immediate Danger: Prognosis for a Radioactive Earth*, The Book Publishing Company, Summertown, TN, 1985, 39.
15. Thompson, T., Facing up to being down, *Esquire*, December 1995, 74.
16. Klerman, G. L. and Weissman, M. M., Increasing rates of depression, *Journal of the American Medical Association*, April 21, 1989, 2229-35.

17. Centers for Disease Control and Prevention, *1995 Assisted Reproductive Technology Success Rates: National Summary and Fertility Clinic Reports*, Atlanta, GA, 1997.
18. Wright, L., Silent sperm, *The New Yorker*, January 15, 1996, 44.
19. Auger, J., et al., Decline in semen quality among fertile men in Paris during the past 20 years, *New England Journal of Medicine*, February 2, 1995, 281-5.
20. Sperm in the news, *Rachel's Environment and Health Weekly*, January 18, 1996, 1-2.
21. Sternglass, E., Evidence for low-level radiation effects on the human embryo and fetus, in Radiation Biology of the Fetal and Juvenile Mammal, *Proceedings of the Ninth Annual Hanford Biology Symposium at Richland Washington*, U.S. Atomic Energy Commission, Washington, DC, December 1969.
22. Prospective Payment Assessment Commission, *Medicare and the American Health Care System: Report to the Congress, June 1997*, Washington, DC, 1997, 135.
23. Gornick, M. E., Thirty years of Medicare: impact on the covered population, *Health Care Financing Review*, Health Care Financing Administration, Baltimore, MD, Winter 1996, 197.

chapter nine

Health effects and the post-1983 generation

The still-unfolding immune health saga of the Baby Boomers, who have lived about half of their expected lives, is becoming clear. Persons born between 1945 and 1965 have often had a poorer health record than the generation just preceding. Many of these downturns in health status have come in areas of immune disease or conditions affected by the immune function. Cancers, low-weight births, infant mortality, common childhood infectious diseases, AIDS, chronic fatigue syndrome, allergies, and sperm concentration are among the most widespread of these. The issue of Baby Boomer health status is not one likely to disappear from the public eye; rather, since the Boomers are now turning 50 and entering the period of life in which disease and death becomes more common, the concern will become more intense. If the Boomer record continues to be substandard, the sheer numbers of this generation entering old age and the enormous costs to society of impaired health — both in direct health costs and loss of able-bodied contributors to society — will be a titanic jolt felt across America.

If this prediction comes to pass, the slippage in Boomer health will become more recognized, and the debate over why this is taking place should sharpen. The objective, insightful research needed to comprehend why these trends occurred has been sadly lacking thus far. But sooner or later, the amount of unexpected suffering by a group of people will compel the nation's people, and thus motivate the nation's leaders, to seek answers, no matter how ugly or unpleasant. Such a strong desire for truth means research will expand its scope into areas like man-made hazards. Radiation exposure is just one of these; food additives, pesticides, herbicides, and industrial pollution are also areas which raise many unexplained questions. Baby Boomers are a generation with a unique set of radiation-related problems. This group was born and raised at a time when atomic weapons were being manufactured and tested, allowing large releases of immune and hormone-damaging radioactive chemicals into the environment. They continued their lives as young adults while over 100 nuclear reactors were churning out small but steady amounts of these products. No other group, born before or since, has the same story, and it is imperative that the research community tell the complete story of what happened to what essentially has been a test of about 75 million guinea pigs.

The Baby Boomers performed worse than the Great Depression/World War II generation and the group born after 1965. "Generation X" had lower infant mortality and low birth weight rates, lower rates of AIDS and CFS, and much lower increases in cancer rates than the Boomers. They are still only young adults, and have much of their lives still ahead of them, but so far, it appears that they are being spared much of the immune-related ravages suffered by the Boomers.

The youngest Americans — infancy

If we roughly classify each generation as a 20-year period, the next group following Generation X would be births between about 1985 and 2005. Many in this population are only babies and children, while the others have yet to be born. However, a number of trends in health status *are stagnating, or even worsening*, compared to the progress made by Generation X, and like the Baby Boomers, many of these reversals are occurring in immune-related conditions. Thus, constructing a health profile of the '80s/'90s generation is warranted.

It is logical to assume that radiation has not affected these youngest Americans like it did the Baby Boomers. The '80s/'90s kids never had to deal with fallout from atmospheric bomb testing or underground testing (the last Nevada test was conducted in September 1992). They were never exposed to the large radioactive releases in nuclear weapons manufacturing such as those in the early years of Hanford and Oak Ridge; by the late 1980s, weapons production had ceased altogether. They haven't experienced (thus far) a serious domestic accident at a nuclear facility, like at Savannah River in 1970 or Three Mile Island in 1979. They were born into an America with over 100 operating nuclear power plants, but with much lower levels of radioactive releases than in the 1970s.[1]

Still, the atomic age is still a relatively new one, and much is yet to be learned about how different types of exposure affects humans. With man-made radiation still in the environment and previously-exposed populations now bearing children of their own, it is important to trace the progress — or lack thereof — in the health of succeeding generations. If this younger generation suffers the same consequences as the Baby Boomers, we need to re-think the assumption that the low levels of radiation exposure in the 1980s and 1990s are not harmful.

Infant mortality

American medicine continues to make remarkable progress in its ability to keep infants alive. The century-long decline, which paused only for the Baby Boomers, continues, to the point where fewer than 1 in 100 babies born today die in the first year of life. Although infant mortality rates for blacks are still more than double those of whites, rates for all races continue to drop. Neonatal treatment is so proficient in keeping babies alive, it is difficult to gauge any effect of a pollutant by analyzing mortality in the first year of life.

Low-weight births

As seen in the analysis of the Baby Boom generation (Chapter 5), birth weight appears to be a sensitive indicator of the effects on infants of environmental threats such as radiation. The low-weight birth debacle of the Baby Boomers (rates increased 2% for whites, 35% for non-whites from 1950 to 1966) was followed by nearly two decades of steady improvement. Between 1966 and 1984, rates dropped 22% for whites and 20% for non-whites. However, progress slowed in the early 1980s, and beginning in 1984, rates turned around and resumed an upward climb; between 1984 and 1995, the rate *increased 11%* for whites and 6% for non-whites. There has been *no progress since 1975* in U.S. birth weights (see Figure 5.2 in Chapter 5). An estimated *300,000* fewer babies in the U.S. would have been born weighing under 2500 grams from 1985 to 1995 if 1966 to 1984 trends had continued. These excess numbers far exceed those experienced in the atmospheric bomb test era.

Over 6%, or more than 280,000 babies each year, are born under 5 1/2 pounds (2500 grams), constituting a serious public health crisis. Of the numerous health goals for the year 2000 set by the U.S. Public Health Service in 1990, low-weight births are one of the few areas to not make any progress, a trend which federal officials have yet to explain. Low-weight babies are at higher risk for birth defects, other medical problems, and death, as described earlier.

More local health departments are now compiling county-specific data on birth weights. This added information makes it possible to analyze effects of living near nuclear plants. One such set of data exists in the area near the Oak Ridge weapons facility in Tennessee, the oldest in the U.S. Although Oak Ridge ceased to make nuclear fuel for weapons in the late 1980s, the plant continued to conduct research on applications of nuclear technology. In addition, the area continued to be confronted with the problem of contamination of the local environment by long-lived emissions from the plant, along with the considerable nuclear waste in temporary storage that has accumulated from the years of operation. The nearby Watts Bar power plant, opened in 1995, will also add radioactive fission products to the local environment in the future. Perhaps most important, the Toxic Substances Control Act incinerator at the Oak Ridge site, which burns low-level radioactive waste from plants across the U.S., opened in 1991, sending emissions into the local atmosphere.

The rate of low-weight births for the Oak Ridge area has always been high during the atomic age. Furthermore, the already high rate rose in the late 1980s and early 1990s. The rate for whites living in the 19 counties less than 50 miles from the plant jumped 16% from 1982 to 1994 (6.51 to 7.55) versus an 11.6% hike for the rest of Tennessee and 8.5% for U.S. whites. Seven downwind counties in southeast Kentucky 50 to 100 miles from Oak Ridge also experienced a 16% increase during this time. The rate for five small mountain counties in western North Carolina, also 50 to 100 miles from Oak Ridge, soared 43%. These counties accounted for *over 1000 excess low-weight births* from the early 1980s to the early 1990s. Effects of the incinerator and continued nuclear research must be considered as potential factors behind this crucial trend.

Fetal deaths

Fetal deaths (gestational age 20 weeks or more) among Baby Boomers were discussed in Chapter 5. As mentioned, this statistic is another indicator measuring the well-being of the infant, but is hampered by under-reporting of cases to local health departments. Nonetheless, the Baby Boomers experienced a lack of progress in fetal deaths, much as they did in infant mortality and low-weight births, while the generation born after 1964 resumed the improvement patterns of before World War II. After 1984, the decline among whites continued, but rates for non-whites stagnated.

Premature births

During the 1940s, 1950s, and 1960s, infant deaths, birth weight, and fetal deaths were among the few indicators of infant health routinely reported by each state to the federal government. These limited statistics for babies born after the early 1980s show trends similar to those of the Baby Boomers. There are some differences: infant mortality rates continue to drop in the 1980s and 1990s, as opposed to stagnating in the 1950s and early 1960s, but post-1983 births seem to be performing worse than Generation X.

Fortunately, by the 1980s mandatory national vital statistics had expanded to include an additional number of measurements of infant health. One such measure of the infant's overall health and ability to thrive is how many pregnancies are carried to full term, defined generally as 37 weeks. Unfortunately, this information appeared in the *Vital Statistics of the United States* annual publications beginning only in 1962; and from 1968 to 1980, a number of states were excluded, limiting good analysis of trends. However, the years 1981 through 1992 can be compared. Rates fell slightly from 1981 to 1984, but beginning in 1984, rates began to rise; between 1984 and 1992, the proportion of white babies born before the 37th week rose from *7.88 to 9.10* per 100 live births (a jump of 15.5%) and that for non-whites went from *15.43 to 16.60* (up 7.6%). This trend is consistent with the rise in low-weight births, described earlier in the chapter; a large proportion of births under 2500 grams are those born after less than 37 weeks gestation.

Premature births may be associated with environmental insults such as direct exposure of the fetus to radiation, or inherited genetic defects from the exposure of either parent. If a fetus' rapidly growing body is unable to divide cells and create new, healthy cells at an adequate pace, its growth may be restricted and it may be unable to sustain itself in the womb. Effects of radiation have been shown to include such impairment of cell division, and should be considered one of the possible factors behind this recent trend.

The prematurity trend could be affected by timeliness of prenatal care. Ideally, prenatal care should begin in the first three months of pregnancy and continue at appropriate intervals through the remaining six months. From 1984 to 1990, the percentage of white American mothers who received no prenatal care or began it only in the third trimester (7th, 8th, or 9th month of pregnancy) rose only from 4.7 to 4.9, while the figure for blacks went from 9.7 to 11.3. However, rates were also rising from 1980 to 1984 (4.3 to 4.7 and 8.9 to 9.7, respectively), during a time when low-weight and premature birth rates were decreasing. While less timely prenatal care may still be a factor, it does not appear that this alone explains why premature births started to increase after 1984.

Apgar scores

Another measure of infant health recently added to national vital statistics tabulations is Apgar scores. Named after Dr. Virginia Apgar, who developed the measure in the 1960s, the score is a summary of five items describing the baby's health (color, heart rate, muscle tone, respiration, and response to stimulation of the sole of the foot). Each of the five receives a score of 0, 1, or 2, so a "perfect" score is 10. Ratings, which are usually assigned by a nurse present at the delivery, are calculated one minute and five minutes after birth.

Apgar scores for all states appear in *Vital Statistics of the United States* beginning in 1978. In general, most babies receive a high score; an average of 8 (at one minute) and 9 (at five minutes) have predominated in the past 15 years. Initial problems that may lead to a score less than 10, such as trouble breathing or yellow skin color, are often resolved by the five-minute mark; about 9% of babies fail to receive a score of 7 or more at one minute, but only 1 to 2% don't get at least a 7 at five minutes.

One unusual and unexplained trend since 1978 has been the dwindling number of babies receiving a perfect 10, at both the one-minute and five-minute marks. More than 37% of 1978 births (white and black) received 10s after five minutes, but only 13% of 1992 births did so. No major change in reporting procedures during this time are known. While the Apgar score is not necessarily the most meaningful measure of a baby's health, and the trend doesn't suggest a collapse in health status, it still must be regarded seriously, especially in light of the many other adverse health trends of babies born in the 1980s and 1990s. As we have seen in previous chapters, radiation and other environmental causes may be linked with subtle changes in health, such as more allergies and slightly lower white blood cell counts.

Congenital hypothyroidism

One condition associated with radiation exposure is hypothyroidism, or underactive thyroid gland, due to inadequate levels of circulating thyroid hormone (known as thyroxine, or T4). Elevated hypothyroid rates after exposure to radioactive iodine, which seeks out the thyroid gland and destroys or impairs thyroid cells, has been documented in certain populations exposed to high doses, such as Hiroshima and Nagasaki survivors and residents of the Marshall Islands.[2,3] The disease is easily detectable through a blood test and controlled by doses of synthetic thyroid hormone.

In the late 1970s, mass newborn hypothyroidism screening became possible after the invention of a simple test using only a drop of the baby's blood. Although the condition is rare in infants, it is easily controlled with synthetic thyroid hormone; if the condition is left untreated, the baby's physical and mental growth will be stunted, because hormones in the thyroid gland are key to childhood development. Years ago, when the disease was not detectable, doctors called the disease "cretinism" to reflect the short, mentally impaired children that often resulted. By 1980, the majority of states had enacted laws to screen all newborns for the disease, and all states were doing so by 1990. While each state's definition of hypothyroidism may differ somewhat by the level of T4, trend analysis of confirmed cases is possible.

From 1977 to 1979, only New England states screened routinely for the disease, and no change in rates from one year to the next was documented (if anything, a slight decrease occurred). For the 30 states (with 64% of U.S. births) providing data every year beginning in 1981, there was little change from 1981 to 1984 in the rate of about 21 to 22 cases per 100,000 births, or just under 1 in every 5000 births. However, beginning in 1984, the rate began to rise and only leveled off after 1992 (Figure 9.1). The astounding jump of 47% (21.17 to 31.16 per 100,000 births) in babies born with hypothyroidism means that 1300 such births now occur each year, instead of about 800 just several years ago. In all but a few states, procedures for reporting confirmed cases of hypothyroidism are the same as they were a decade ago. Testing methods are the same. A national system of reporting that began in 1988 may have encouraged more thorough reporting of cases, but this should be only a one-time effect; instead, the rate continued its steady ascent year after year. This trend should be seen by analysts of the effects of radiation as one of the more serious patterns in newborn health, as it is linked directly with a radioactive chemical such as iodine. Even though hypothyroidism is rare and only affects a small number of babies each year, we must once again consider that in addition to those meeting the criteria for illness, there are also millions more suffering from more subtle outcomes of radiation exposure. In other words, the rising hypothyroid rate may be accompanied by a slight decrease in thyroid hormone for many other babies, impairing their physical and mental growth somewhat, but falling short of meeting the

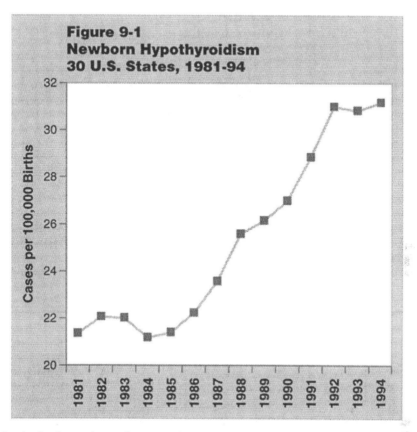

Figure 9-1
Newborn Hypothyroidism
30 U.S. States, 1981-94

criteria for hypothyroidism. Unfortunately, there are no known studies on trends of average thyroid hormone levels in newborns.

Abnormal birth conditions, congenital anomalies

One significant addition to national vital statistics reporting happened in 1989, when detailed information on the numbers and types of birth defects, or congenital anomalies, was reported. Unfortunately, with only four years of data available thus far, no significant trend analysis is possible. However, the data are helpful in understanding the extent of birth defects, and in pointing out how premature births/low-weight births are strongly associated with a greater risk of birth defects. The current number of birth defects in newborns is 50,000, or just over 1% of the nearly 4,000,000 births each year. The most common of these are heart malformations, abnormal numbers of fingers and toes, cleft lip or palate, malformed genitalia, club foot, Down's syndrome, and hydrocephalus.

Several states that began registries of birth defects in the 1980s now have nearly 10 years of data to analyze. One of these states is New Jersey, which released 1985–93 data in April 1996. The rate of congenital birth defects increased by a surprisingly large 45% during the eight-year span, with the 1993 rate standing at 3.5 per 100 births. The annual number of birth defects in New Jersey soared from *2514 to 4092*. The rate in Ocean County, home of the safety violations-plagued Oyster Creek nuclear power plant, rose 52% and now has the highest birth defects rate of any New Jersey county not located near the New York City metropolitan area. In California, the state's Birth Defects Monitoring Program showed rates rising from 2.7 to 3.2 per 100 births from 1983 to 1987, only to fall to 2.6 by 1990.

One birth defect that is not very common but extremely debilitating is fetal alcohol syndrome. In 1995, the U.S. Centers for Disease Control and Prevention (CDC) published a study in its *Morbidity and Mortality Weekly Report* on the rising incidence of the disease. Beginning in 1983, the disease started to become much more common in newborns; by 1993, the rate of 6.7 per 100,000 stood more than five times greater than the early 1980s (1.2), and far above the national goal for the year 2000 of 1.2. The CDC authors suggest that the increase could be due to greater awareness of the disease or improved coding practices, but they stop short of attributing all of the increase to these factors, leaving other causes, including environmental factors, as potential contributors.

Congenital AIDS cases

Because the AIDS virus is usually transferred by blood and semen, much of the epidemic is confined to the adult population. However, the virus also can be transferred from mother to child during pregnancy, and because 25% of infected mothers pass the virus to their children, about 1500 to 2000 unfortunate babies were born with the virus each year in the mid-1980s and early 1990s.[4] From 76 cases in 1983, 936 AIDS cases were diagnosed in children under 14 in 1992, representing congenital AIDS cases. A total of 7902 pediatric AIDS cases had been diagnosed by June 30, 1997, with 58% of these already dead from the disease and another estimated 3500 children carrying the virus but not yet sick with any infectious illness that is the mark of full-blown AIDS. This trend certainly reflects the greater levels of risk among mothers but also may represent a more susceptible infant; not much is known about why some people exposed to the virus develop the disease and others don't.

One of the bright lights in a largely dismal record of AIDS treatment in the past 15 years concerns the control of congenital AIDS cases. If HIV is detected in a pregnant woman and a combination of drugs featuring the antiviral medication zidovudine, or AZT, is administered during pregnancy, about two thirds of otherwise HIV-positive newborns will be born without the virus. Already, there are signs that childhood AIDS is receding; the 1992 high of 936 newly diagnosed cases under 14 years of age fell to 881 in 1993 and 769 in 1994. In addition, experts believe congenital AIDS increases are slowing because HIV-positive women are fewer in number, less fertile, or are having more abortions.[5]

Diarrheal deaths in infancy

In 1995, the *Journal of the American Medical Association* ran an article on trends in deaths among infants due to diarrhea.[6] Mortality due to this condition was once common among infants in the U.S., and still is not a rare occurrence in third-world countries. Because of various therapies now in use in the U.S., the annual number of infants under one year dying from diarrhea-related conditions fell sharply each year since the 1960s, declining from 1112 to 240 between 1968 and 1985. However, the rate stabilized from 1985 to 1991, the last year studied. This lack of progress should be considered in light of the trends in many other indicators of infant health covered in this chapter. In addition, the article states that diarrhea-related disease is linked to either viruses or bacteria (such as salmonella and *E. coli*), meaning that whether or not the baby develops the disease or dies depends on the strength of the child's immune system to fight off these invaders. After radiation-related immune suppression, one would expect certain viruses and bacteria to affect more people and become more lethal.

Childhood

Childhood deaths, all causes

As babies born in the 1980s and 1990s moved out of infancy into childhood, an unexpected slowing of progress in mortality rates occurred. Children age 1 to 4 have traditionally had a much lower death rate than babies in the first year of life; the current rate for young children is about 5% of the infant rate.

The annual death rate for children age 1 to 4 (minus accidents and homicides) was cut nearly in half from 1970 to 1985, falling from 51.1 to 29.0 per 100,000 persons (Figure 9.2). However, the pre-1985 annual decrease of 5% had slowed to just 1% from 1985 to 1991, before dropping sharply in 1992. An excess of several thousand young American children died from 1985 to 1991 for reasons thus far unknown.

Childhood cancer incidence, U.S.

Although babies born after the mid-1980s are still children, many have moved past infancy, and a child health profile is emerging. Any assessment of possible effects of radiation exposure on children must include cancer. In the Baby Boom years, cancer incidence could only be calculated by using Connecticut's tumor registry. By the 1970s, however, national figures became available. When President Nixon declared the War on Cancer and Congress passed the National Cancer Act in 1971, one of the actions that ensued was the publication of an annual compilation of cancer incidence statistics. The calculations used data from five states (Connecticut, Hawaii, Iowa, New Mexico, and Utah) and four metropolitan areas (Atlanta, Detroit, San Francisco/Oakland, and Seattle), each of which had an established cancer registry. These

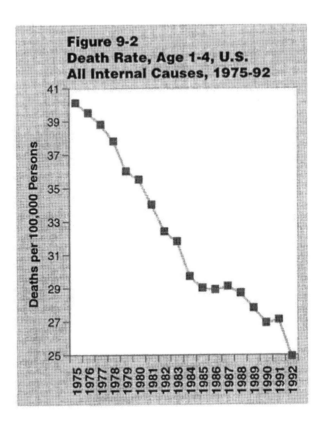

Figure 9-2
Death Rate, Age 1-4, U.S.
All Internal Causes, 1975-92

nine locations represent over 20 million Americans, or just under 10% of the population, making them a reasonably representative sample of the country. This new national cancer registry, run by the National Cancer Institute, is known as the Surveillance, Epidemiology, and End Results (SEER) program.

From 1973 to 1980, SEER data show that there was little change in childhood cancer rates. After 1980, however, all-cancer and leukemia rates increased. Adding data from three areas with cancer registries established by 1980 (Denver, New York, and Wisconsin) to the nine SEER regions creates a database including about 19% of U.S. children and enables a trend analysis from 1980 to 1993 to be made.

Figure 9.3 shows that an unchanged cancer rate for children under 10 from 1980 to 1982 was replaced by a steady rise until 1993, the last year for which complete data are available. *Incidence climbed 37% from 1980–82 to 1991–93.* If these states and cities are typical of the entire U.S., the annual number of cancers 0 to 9 has jumped from just over 4000 to about 6000, meaning a total of about *10,000* excess childhood cancers occurred in the U.S. from 1983 to 1993.

Several interesting findings are embedded in this trend. Cancer increases are greatest in the youngest children, i.e., the rate of new cases for children under one year rose 54%, compared to those 5 to 9 (30%). While the cancer

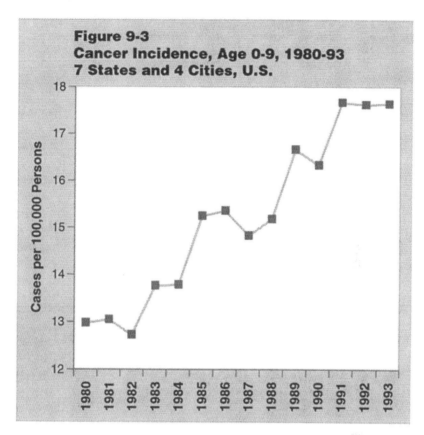

Figure 9-3
Cancer Incidence, Age 0-9, 1980-93
7 States and 4 Cities, U.S.

rate rose in all areas studied, the strongest advances occurred in the Detroit metropolitan area and New York state. (Perhaps not coincidentally, Detroit is 30 miles from the Fermi nuclear facility, while the New York City area is about 35 miles from the Indian Point nuclear plant; both of these experienced considerable safety problems during the 1980s.) The increase for the three non-SEER areas (48%) far exceeded that in the eight SEER regions (25%). Thus, we learn that limiting analyses only to SEER areas may not represent the entire nation.

Possibly the most important finding in the data is the large rise in cancer rates for persons born in 1986 and 1987. These children received *in utero* exposure to the Chernobyl fallout. Children born after 1987, who would only receive some exposure from Chernobyl's long-lived radionuclides had childhood cancer rates similar to the 1986–87 birth cohort.

Another question posed by this childhood cancer outbreak is what types of cancers account for this trend. Childhood cancers can generally be divided into three types (leukemia, brain/central nervous system cancer, and all other cancers) each accounting for about one third of the total. Leukemia is a type of cancer associated with radiation exposure, especially in children. From the early 1980s to the early 1990s, leukemia incidence in those 0 to 9 rose 15% in the 11 states and cities. This increase falls short of the 37% mark for all cancers, so other malignancies in addition to leukemia are contributing to the current childhood cancer epidemic in the U.S.

Childhood cancer near nuclear plants

Unfortunately, the National Cancer Institute's study of cancer in counties near nuclear plants stops in 1984, leaving the story of the youngest generation of Americans an incomplete one. Still, comparing childhood cancer incidence data from the early 1970s to the early 1980s demonstrates unusual increases. For all cancers except leukemia, incidence for children age 0 to 9 in the home counties of Connecticut's Haddam Neck and Millstone plants and Iowa's Duane Arnold plant soared from 29% *below* to 37% *above* the state rate. For leukemia in those under 10 years old, the rate near the four plants jumped from 35% *above* to 67% *above* state averages. While these data are only based on 61 cases in the early 1980s, they still give a hint that one characteristic of childhood cancer increases beginning in the early 1980s involve youngsters living near nuclear power plants.

The negative trend in childhood cancer has not escaped detection. In a 1991 *New York Times* article, renowned researcher Devra Lee Davis said, "This is a troubling trend. Even if we succeed in curing childhood cancers at an increased rate, at the pace we're going we'll soon have one in a thousand adults being a survivor of childhood cancer, and these people remain at risk for other cancers and illnesses. There are real costs to the treatment."[7] Thus, regardless of the causal factors, the toll of excess childhood cancers is mounting, in terms of deaths, suffering, and economic losses.

Childhood cancer mortality, U.S.

In the last half century, advances in diagnosis and treatment have lowered childhood cancer mortality rates considerably, despite the growing number of cases. Current rates are only about one fourth of what they were in the mid-20th century, even though about 1700 children under 15 still lose their lives to cancer each year.

Still, the 1980s have produced some abrupt halts in mortality drops for children under 15. No reduction occurred for children 0 to 4 (1980 to 1983), age 5 to 9 (1984 to 1987), and 10 to 14 (1987 to 1990), a drastic abnormality to the steady, decades-long pattern of decreases. The vast majority of children in these groups were born in the years 1978 to 1982. It will be interesting to see if those born after 1982 will continue this pattern of diminishing improvements.

Childhood cancer mortality near nuclear plants

Although the NCI collected cancer incidence data from only nuclear counties in Connecticut and Iowa, it included mortality statistics for counties near all power plants. Another way to evaluate the effects of radiation on children is to look at cancer deaths for children under 10 for counties nearest nuclear sites. A comparison was made of childhood cancer rates near 13 plants that began operations from 1974 to 1976 from the most recent period before the plants opened (early 1970s) to the time 6 to 10 years after operations began (early 1980s). The combined 13 nuclear plants, with a total of 21 reactors (some sites have more than one) allow us to analyze 25 home counties in which over 500,000 children under 10 reside.

The results show powerful evidence linking nuclear emissions with childhood cancer. For all cancers except leukemia, the death rate among those 0 to 9 in the 25 counties roared from *30% below* to *14% above* the U.S. rate between the early 1970s and early 1980s. Leukemia also increased, but not nearly as much (4% below to 3% above). A total of 132 cancer and leukemia deaths in those age 0 to 9 makes this analysis a meaningful one, not just based on a handful of deaths. Two of the most densely populated areas near the nuclear plants saw their cancer and leukemia death rates *double* (compared to U.S. rates) in 10 years. One is the Rancho Seco site near Sacramento California, which operated for only 15 years before closing in 1989 (more on Rancho Seco's health effects is presented in Chapter 10). The other is the Three Mile Island plant near Harrisburg, Pennsylvania, also opened in 1974, and best known for the partial meltdown of its reactor core in March 1979 (Chapter 4). Any debate on whether the accident affected the health of local residents becomes more one-sided after considering these subsequent death rates from childhood cancers, along with the infant mortality rises presented in Chapter 4.

The significance of the childhood cancer mortality increases in the 1980s near nuclear plants cannot be understated. These increases only represent counties nearest (within 30 miles) the reactor; the damage may well extend to persons living beyond the 30-mile mark. Moreover, several dozen reactors opened in the 1970s and 1980s. These machines, much larger than their predecessor models, were also touted by manufacturers and utilities to be much safer. The new reactors relied on pressurized water, compared to the earlier generation of boiling water reactors, to generate electricity. However, the promise of a safe group of reactors seems to have all but vanished. The notorious Three Mile Island reactor ran on pressurized water. In the mid-1990s, utilities operating 14 pressurized water reactors in the Midwest filed suit against their manufacturer, Westinghouse Industries, charging that faulty design led to pipe corrosion and leakages well before the expected lifespan of 40 years. The suits have still not been settled, but during the first decade of the 21st century, with these 1970s-era plants turning 30 years old, safety and health will play a key role in whether or not operators and regulators decide to retain or scuttle these reactors.

Asthma

Another health problem plaguing society in the 1980s and 1990s that has affected the youngest group of Americans is the growing presence of asthma. Because asthma is a type of allergy, the body's immune system strength is crucial in whether a person develops the disease; and an environmental

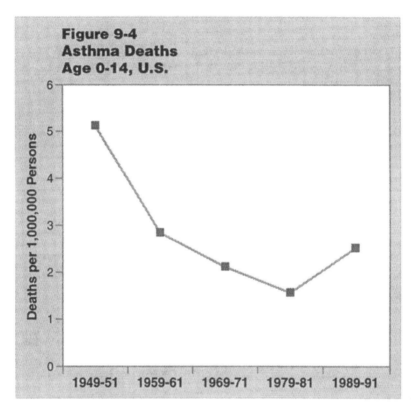

Figure 9-4
Asthma Deaths
Age 0-14, U.S.

contaminant (such as radiation) affecting the immune system may be a risk factor. As mentioned previously, incidence rates are rising in recent years Because asthma attacks sometimes require hospitalization, admission rates are rising. Between 1980 and 1992, admissions for asthma for children under five *rose 60%*, from 38 to 61 per 10,000 persons,[8] at a time when hospitaliza tions for most other conditions were falling. This increase translates into ar excess of *45,000* hospital admissions for asthma in small children each year or about $80 million in inpatient hospital costs.

While few children die from asthma, death rates still dropped sharply throughout the 1950s, 1960s, and 1970s, falling a total of 70%. However beginning in the late 1970s, a turnaround took place, and by 1990, the rate had risen 60%, from 1.5 to 2.5 deaths per million population (Figure 9.4) The number of American children under 15 who die from asthma has jumpec from 80 a year in the late 1970s to the current 135 per year. It may not be coincidental that both hospitalizations and deaths for childhood asthma went up 60% each in the past decade.

Pneumonia

While the childhood death rate due to pneumonia continues to fall, the disease may be more widespread than a decade ago. Between 1985 and 1995 the number of annual hospitalizations for the condition for children 0 to 14 went from 269,000 to 383,000; factoring in an increase in the population, the hospitalization rate *rose* 27%. Even though these children are often discharged home within five days, an extra *$200 million* is spent each year to treat them in hospitals, as compared to the mid-1980s.

Bronchitis

Another respiratory disease that affects many children is bronchitis, a swelling of the lungs that causes wheezing and coughing. Although most cases are readily treated and don't last long, some cases in children are chronic or frequently recurring; and some result in severe debility or death. Similar to asthma, death rates for bronchitis in children were falling sharply until recent years. From 1979–81 to 1989–91, the rate doubled, from 1.722 to 3.656 per 1,000,000 Americans under 15. While only 70 American children die each year from bronchitis, the trend in death rates may mirror patterns for incidence and severity of the disease, which affects hundreds of thousands of children.

Septicemia

Septicemia, or blood poisoning, was shown in Chapter 8 to be increasingly fatal to Baby Boomers as they move into middle age. Although fatalities from this disease often involve the elderly, septicemia also kills 400 U.S. children under 15 each year. We know that the septicemia death rate for children under 15 rose 55% in the late 1940s and 1950s (Baby Boomers) and fell in the 1960s and 1970s (Generation X). By 1989–91, the rate had inched up 2% from a decade earlier, halting the progress of the post-Boomers.

Communicable childhood diseases

Chapter 6 demonstrated that a number of unusual increases among Baby Boomers took place for diseases such as measles and scarlet fever, for which public reporting is required. Trends for these disorders often reflect the presence of epidemics. However, when many of these conditions become more common simultaneously, the ability of the at-risk population to fight off disease should be considered.

Six of the reportable diseases usually strike in early childhood. The most frequently occurring of these is chicken pox, or varicella, with over 150,000 reported cases a year, 80% of which are in those under 10 years old (Table 9.1). There was a slight increase in chicken pox during the 1980s, before a decline in the early 1990s. Future chicken pox rates will probably drop after the introduction of an effective and safe vaccination in 1995.

However, reported cases of the other five diseases increased in the 1980s and 1990s, as the post-1980 generation reached the age at which these diseases occur. Over half of the victims of these diseases are age 0 to 9. Aseptic

Table 9.1   Number of Cases, Selected Reportable Conditions, U.S., 1970–94

Period	Aseptic Meningitis	Chicken Pox	Measles	Mumps	Whooping Cough	Shigellosis
1970–74	24,333		203,700	432,847	14,733	95,437
1975–79	28,101	879,804	176,819	150,617	8,611	85,422
1980–84	48,277	958,528	22,428	25,163	9,612	94,119
1985–89	50,988	952,899	34,348	34,198	18,214	113,682
1990–94	60,381	764,480	40,941	15,357	22,575	136,523

meningitis more than doubled from 1975–79 to 1990–94, as did pertussis/whooping cough. Shigellosis, the second most common childhood disease behind chicken pox, rose 60%, from 17,000 to 27,000 cases a year. Measles cases jumped 80% from the early 1980s to early 1990s. Again, these unusual trends may reflect the presence of epidemics, but also may point to the susceptibility of the at-risk population.

Childhood ear infections

Children are more susceptible than adults to infections of their middle and inner ears, causing them discomfort and pain. These disorders are typically short-lived viral and bacterial conditions which are easily treated by antibiotics or even bed rest. Although there is no requirement for reporting ear infections to public health departments, periodic surveys estimate how often they occur. From 1975 to 1990, the rate of childhood ear infections nearly tripled (*increased 175%*), according to the National Center for Health Statistics. A related condition, sinus infections, more than doubled between 1985 and 1992.

Some health professionals are concerned that antibiotics are overused for conditions like ear infections and that a course of "watchful waiting" and bed rest might be just as effective in many cases. Extensive use of antibiotics may cause resistance to the drugs and allow infections to flourish. Another possibility that must be considered is that such a massive increase in ear infections may reflect the population's inability to resist disease, even before medication is taken. When contagious bacteria or viruses circulate through households, schools, day care centers, and playgrounds, some children will become ill and some will remain healthy, largely due to the strength of the immune response.

Obesity

Another disease on the rise among young children born in the 1980s and after is obesity. The proportion of *overweight children more than doubled*, from 5.0% to 10.6%, from 1963–65 to 1988–94. Most of the increase occurred during the 1980s and 1990s. "Overweight" is defined as exceeding the 95th percentile of the normal range of weight based on a child's age, sex, and height.[9]

Obesity is not a disease per se, but research has shown that heavy children have a greater risk later on in life for gallbladder disease, osteoarthritis, diabetes, heart disease, some cancers, and early death. Obesity is seen as a function of overeating, eating the wrong foods, and a sedentary lifestyle. Increasingly, however, clinicians are finding that there may be a genetic predisposition to being overweight, and we know that genetic damage can be caused by radiation. Furthermore, it is well established that radioactive iodine, which seeks out and disrupts the thyroid gland, has an effect on metabolism. When large numbers of thyroid cells are damaged, it is more likely to reduce a person's vitality and make them sluggish and sedentary, and unable to utilize the body's energy supply from food intake. Rosalie Bertell says that "a mild exposure experienced by a large population could cause a decrease in average thyroid hormone levels and an increase in average body weight."[10] A mild exposure is precisely what has happened to American children in the 1980s, as iodine emissions from nuclear power plants continue.

Nuclear accidents

Another insult to the post-1980 generation is the Chernobyl accident of April 26, 1986. The immediate victims of the accident were the excess number of infant deaths and congenital anomaly deaths from May to August 1986, discussed in Chapter 4. By the mid-1990s, other afflictions hitting young Americans in unexpectedly high numbers began to appear in the medical literature:

- Thyroid cancer for American children age 0 to 14 rose 13% in the early 1990s. The period of years after 1990 was also the time when much larger increases were found in children of Belarus and the Ukraine, near the Chernobyl reactor, who absorbed massive doses of radioactive chemicals.[11] Although thyroid cancer diagnosed in children is rare, the 13% increase approximates the rise for thyroid cancer in American adults as well in the same period.
- Leukemia diagnosed in the first year of life for U.S. children born in 1986 or 1987, who were exposed to low-level Chernobyl fallout as fetuses, was 30% greater than those Americans born in all other years of the 1980s. A rise of 160% in Greek infants[12] and 48% in West German infants,[13] who were exposed to about 100 times more radioactivity from Chernobyl than were Americans, had previously been documented.
- Hypothyroidism in U.S. newborns rose in 1986–87, continuing a trend begun in 1984. However, the 1986–87 increase was highest in Pacific northwest states (+23.3%) where post-Chernobyl fallout was greatest. The southeastern states, with the lowest fallout levels in 1986–87, saw its newborn hypothyroid rates actually fall 1%.[14] A final, ominous warning about what Chernobyl may have done to American children is the rise in cancer incidence in children born in 1986 and 1987.

Closer to home, a number of accidents at nuclear plants occurred in the 1980s and 1990s, although none was anywhere close to the magnitude of the Chernobyl meltdown. We have already examined the effects of the Three Mile Island accident on local residents, including infants and children. However, there are other accidents that may have caused harm to the local infant and child population. One example occurred at the Plymouth (MA) Nuclear Power Station in June 1982. At that time, officials at Plymouth, a boiling water reactor in operation since 1972 with a checkered safety record, sent word to the Nuclear Regulatory Commission that two accidents occurred early in the month, resulting in the release of radioactive resin beads, fine particulates from waste, and other radioactive products into the local environment.[15] Because of this and other problems, the NRC fined the plant's operator, the Boston Edison Company, a total of $550,000, the largest NRC fine in its history to that date.

Ernest Sternglass wrote an unpublished paper documenting some of the adverse health trends in the area for persons all ages. Infant health can also be analyzed for the state of Massachusetts, since over 60% of the population resides within 50 miles of the plant (Plymouth lies 25 miles southeast of Boston). A comparison of the years 1981 and 1982 is perhaps the easiest way to examine the accidents' impact, even though only the final seven months of 1982 are affected and any adverse trend may be understated (Table 9.2). A total of 68,575 white and 5490 black babies were born in the state in 1982.

For each of the indicators of newborn health listed in Table 9.2, rates in Massachusetts jumped from 1981 to 1982, while U.S. rates either edged up at a slower pace or fell. Both white and black babies suffered from this decline at a roughly equal rate. The indicators with the greatest excesses in Massachusetts were babies born under 3½ pounds, deaths from birth defects, Apgar scores, and cases of hypothyroidism. These measures are not as high-profile as infant deaths, and a number of these babies might survive due to the heroic efforts of medical science. Larger excesses in these conditions lend further support to the idea that low-level radiation exposure has subtle effects that allow people to live, rather than killing its victims instantly, Hiroshima-style. However, these survivors are more likely to have mentally and physically impaired growth, be at risk for certain diseases later in life, and have a shorter life expectancy. Accidents like the ones at Plymouth may well play a role in the health decline of the population born after the early 1980s.

Summary

The group of Americans born after 1983 and continuing into the mid-1990s inherited an improving record of immune disease from their Generation X predecessors (born in the late 1960s and 1970s). However, the positive trends in immune disease among American infants and children began to backtrack, a transpiration similar to what had happened to the Baby Boomers in their

Table 9.2 Changes in Infant Health, Massachusetts vs. U.S., 1981–82

Indicator	Cases in Mass. 1982	% Rate Change Mass.	% Rate Change U.S.
Deaths <1 Year			
Whites	658	+3.2	−3.8
Blacks	102	+16.3	−2.0
Deaths <28 Days			
Whites	495	+7.5	−4.2
Blacks	75	+24.5	−2.2
Fetal Deaths >20 Weeks Gestation			
Whites	490	+2.9	−1.2
Blacks	87	+8.2	−0.8
Deaths <1 Year from Birth Defects			
Whites	172	+20.3	+0.8
Blacks	14	+7.7	+3.1
Births <1500 g (3½ lbs.)			
Whites	669	+8.9	+1.4
Blacks	154	+29.0	+1.7
1-Minute Apgar Score 0–1*			
Whites	539	+26.8	+2.1
Blacks	129	+47.1	+4.1
5-Minute Apgar Score 0–1*			
Whites	178	+18.6	+0.2
Blacks	38	+26.7	+3.1
Newborn Hypothyroid Cases			
All races	24	+17.1	+3.3

* Apgar scores (0 to 10) measure ability to thrive 1 and 5 minutes after birth.

youth. Underweight births rose. Cancer incidence advanced. Overall death rates stopped declining. Cancer mortality near nuclear plants rose. Asthma hospitalizations and deaths skyrocketed, as did bronchitis deaths and cases of ear infections.

These trends are, perhaps not coincidentally, a repeat of what happened to the Baby Boomers 40 years earlier. Even though Boomers were exposed to considerably greater levels of radioactivity *in utero* and during infancy than the post-1983 Americans, the trends for the two groups are remarkably similar. Radiation exposure is most likely affecting these 1980s/1990s Americans, as evidenced by post-Chernobyl disease increases and excess disease in children living near nuclear installations. Once again, it is likely that there is much more to learn about this relatively new technology, now just turning 50 years old. Perhaps the Petkau theory, stating that radiation's effects are most pronounced *per dose* at lowest levels of exposure, is part of the explanation for the recent decline in child health status in America. Strange though it may seem, health effects of emissions from over 100 large nuclear reactors generating electricity may eventually rival those caused by large-scale atmospheric bomb testing and bomb production.

References

1. Tichler, J., et al., *Radioactive Materials Released from Nuclear Power Plants: Annual Report 1993*, Brookhaven National Laboratory, Upton, NY, 10.
2. Nagataki, S., Thyroid diseases among atomic bomb survivors in Nagasaki, *Journal of the American Medical Association*, August 3, 1994, 364-70.
3. Conard, R. A., Dobyns, B. M., and Sutow, W. W., Thyroid neoplasia as late effect of exposure to radioactive iodine in fallout, *Journal of the American Medical Association*, October 12, 1970, 316-24.
4. *The New York Times*, February 21, 1994, A1.
5. *The New York Times*, September 27, 1995, C9.
6. Kilgore, P. E., Holman, R. C., Clarke, M. J., and Glass, R. I., Trends of diarrheal disease-associated mortality in U.S. children, 1968 through 1991, *Journal of the American Medical Association*, October 11, 1995, 1143-8.
7. *The New York Times*, June 26, 1991, D22.
8. Centers for Disease Control and Prevention, Asthma mortality and hospitalization among children and young adults — United States, 1980–1993, *Journal of the American Medical Association*, May 22/29, 1996, 1535-6.
9. National Center for Health Statistics, *Health United States 1996–97 and Injury Chartbook*, DHHS Publication (PHS), 97-1232, Hyattsville, MD, 1997, 193.
10. Bertell, R., *No Immediate Danger: Prognosis for a Radioactive Earth*, The Book Publishing Company, Summertown, TN, 1985, 39.
11. Stsjakhko, V. A., Tysb, A. F., Tronko, N. D., et al., Childhood thyroid cancer since accident at Chernobyl, *British Medical Journal*, March 25, 1995, 801.
12. Petridou, E., et al., Infant leukaemia after *in utero* exposure to radiation from Chernobyl, *Nature*, July 25, 1996, 352-3.
13. Michaelis, J., et al., Infant leukaemia after the Chernobyl accident, *Nature*, May 15, 1997, 246.
14. Mangano, J., Chernobyl and hypothyroidism, *Lancet*, May 25, 1996, 1482-3.
15. June 4, 1982 letter from R. D. Machon of Pilgrim (MA) Nuclear Power Station to the Nuclear Regulatory Commission Region I, King of Prussia, PA.

part three

---

*The great challenge for present
and future America*

chapter ten

Demonstrating low-level radiation's health effects: convincing scientists and the public

It may be repetitive to say again, in light of the enormous evidence presented in the previous chapters, but there is a statistical link between low-level radiation exposure and many immune-related health effects on the Baby Boom generation. The link began when Boomers were fetuses, infants, and children, and has continued into middle-age. In addition to the Boomers, the generation of children born after about 1983 appears to be experiencing the same adverse immune-related health trends as the Boomers did. These linkages have been repeatedly shown not just for all Americans, but for those living near nuclear plants and those exposed to high amounts of fallout from atomic bomb testing. Therefore, even though national trends are influenced by many factors, environmental and nonenvironmental, identifying excess disease and death near nuclear plants clearly implicates radiation exposure as one of these causes.

Means of demonstration

The evidence is powerful, but does this book constitute proof that there is a cause-and-effect linkage between ingesting radioactivity and higher rates of immune disease? While the book alone does not constitute complete proof, it provides a very crucial piece of such a demonstration. A true proof that a substance has harmful effects on humans, which I define as one that satisfies the majority of the scientific community and the public as a whole, is elusive and often takes a long time. The task is especially arduous for low-level radiation, since the technology is only a half-century old and because research in this field has been hampered by political obstacles, discussed in Chapters 2 through 4. Steps that must be taken to achieve proof are as follows.

Beliefs. A hypothesis proposing a cause-and-effect of a harmful substance on human health must exist before any assessment can be made. At the dawn of the atomic age, even the most objective scientists did not believe such a connection was possible. Everyone acknowledged that atomic weapons were vastly more powerful than any means of destruction that man had developed previously, but took no notice of any effects caused by exposure to low levels of these same radioactive chemicals in the bombs. Perhaps the terrible effect of the bomb was all-engrossing. Perhaps the initial research showing effects on Hiroshima and Nagasaki survivors, indicating low levels were safe, was soothing to scientists and the public. Or perhaps the government's all-out commitment to the nuclear arms race blocked any chance of objective research on the effects of low-level exposure. Eventually, however, alternative theories began to surface. Alice Stewart's landmark work in 1956 and 1958 on childhood cancers in those exposed to pelvic X-rays *in utero* was the first to link low-level radiation to increased risk of disease. Abram Petkau stretched Stewart's thesis in 1971; after extensive lab work, Petkau asserted that low-dose exposures, ingested over a prolonged period of time, are more effective *per dose* at breaking cell membranes than single, high-level exposures like Hiroshima and Nagasaki. Beliefs that low-level exposure was harmful were also accepted by segments of the public. Utah sheep herders were not satisfied with the government's explanation that thousands of unexpected sheep deaths were due to malnutrition rather than fallout from a recent Nevada bomb test. Public concerns about bomb testing were strong enough to encourage Adlai Stevenson to propose a curb on tests when he ran for President in 1956.

Anecdotes. Anecdotal evidence supporting a theory is often the next step in proving a cause-and-effect linkage. Sometimes, anecdotes precede the hypothesis: when five young gay men in Los Angeles died in 1981, the theory that a new disease had surfaced was immediately considered. In the case of low-level radiation's effects, anecdotes probably came before the theory as well. In the 1950s, in the midst of a furious nuclear arms race marked by the detonation of dozens of atomic weapons in the U.S. and Soviet Union, concerned scientists like Linus Pauling pointed out that deaths in children from leukemia were rising noticeably, and charged that these deaths were due to ingesting food and drink contaminated by airborne bomb fallout brought into the food chain by rain and snow. As shown in previous chapters, the anecdotes continued, often in areas hard-hit by bomb test fallout or areas close to nuclear weapons or power plants, raising the cry to better understand any radiation-immune disease.

Statistical Studies. Before a cause-and-effect of a harmful substance is established, epidemiological studies involving statistical links between the offending substances and health effects are required. Numerous journal articles, first begun in the 1960s by Ernest Sternglass, have indicated such a link between low-level radiation and immune-related disease. This book has explored the same connection for many diseases for a single population, the

Baby Boomers (and later, the post-1983 generation). Statistical studies are powerful tools which often guide clinicians performing the followup medical research. However, helpful they are, statistics *by themselves* often cannot prove a cause-and-effect, often because there are other potential factors that can affect disease and death trends. There are exceptions to this rule, the most famous being cigarette smoking and lung disease. For years, scientists and the lay public had observed smokers develop a variety of lung disorders at unusually high rates. Many studies were published that *statistically* linked smoking to a higher risk for lung disease, but failed to establish exactly how nicotine and other hazardous substances caused sickness. After the famous U.S. Surgeon General's report in January 1964, virtually all professionals and lay persons were convinced that smoking caused lung cancer and other respiratory disease, leading to aggressive public health actions to reduce smoking.

Clinical Studies.   The next logical step after statistical connections are established is to conduct clinical research. This type of study may involve administering the suspected substance to lab animals in varying doses, and comparing health outcomes with animals receiving no dose (known as a case-control study). In addition, researchers may make measurements of amounts of substances in individuals, and analyze disease rates or indicators of health status, according to the amount of the substance present. This enables a precise *dose-response* analysis to be made, demonstrating how much of the toxic substance is needed to cause *x* number of cancers and other diseases. In the field of low-level radiation, this process can be quite laborious, time consuming, and naturally, expensive.

Scientific/Public Acceptance.   Once both statistical and clinical studies have been completed, evaluation of any cause-and-effect can begin. The new information is analyzed by medical researchers, educators, and practitioners; sometimes, there is a consensus between these parties, and the new information makes its way into medical textbooks, medical education courses, and clinical guidelines used by doctors. Sometimes, there is a conflict within the scientific/medical establishment; in the 1950s, Alfred Sabin and Jonas Salk both developed seemingly helpful vaccines against the dreaded polio virus. After some years of discussion and fighting between the two factions, Sabin's live-virus vaccine became accepted as the safer and more preferable of the two. The public also needs to be informed of a new scientific principle, and many Americans have enough trust in science and medicine to accept what their doctor tells them. However, in some cases like a potentially toxic substance causing human health to be impaired, the lay public may often be ahead of science, suspecting that the substance is harmful even though full "proof" has not been demonstrated. This appears to be the case with low-level radiation; many Americans are leery of living near nuclear reactors and fear nuclear accidents, even though the government and nuclear industry continue to profess that no potential harm is possible.

In the mid-1980s, I lacked any prior convictions about low-level radiation and it's potential health effects. Then I was struck by chronic fatigue syndrome, a debilitating long-term condition that undid my life by taking my ability to hold a job, earn an income, maintain a household, and conduct a social life. Knowing that immune system dysfunction was central to the disease, I became fascinated with potential causes of the condition. I also noticed that some pioneering doctors were suggesting that an environmental component could well be behind the ascent of CFS in the 1980s, and that many of its victims were young adults like myself, born in the Baby Boom era. In late 1988, as I was slowly emerging from the depths of sickness and debility, I read an article in my father's *AMA News*, the weekly newspaper of the American Medical Association, by scientists Jay Gould and Ernest Sternglass, demonstrating that in the four months after Chernobyl death rates in the U.S. had increased by the greatest amount of all time. I had been struck with CFS in December 1986, just eight months after the explosion, a time when an apparently large rise in the condition attracted considerable media attention. Within months, I joined the New York-based Radiation and Public Health Project with which Gould and Sternglass were affiliated, offering my skills as a public health administrator, particularly health policy and epidemiology.

After a great amount of reading and contemplating, I gradually became convinced that Chernobyl and low-level radiation exposure were among the factors — there probably are more than one — behind the CFS epidemic. And I also became a believer, after poring through dozens of books and studies, in the theory that low-level radiation contributed to the rise in numerous immune-related diseases, particularly in the Baby Boomers, who were taking the brunt of the immune scourges of CFS, AIDS, and increased cases of cancer. Although I was doing all this work for no pay at a time when I had little money, I was eager to contribute information to the radiation debate. I saw that two tasks were necessary to advance the understanding of the effects of low-level radiation: convincing the scientific community and convincing the lay public. This two-way effort has taken me down several roads during the past decade.

General statistical studies — Oak Ridge

I quickly learned that sometimes a research effort is a well-planned product, while other times the concept hits you quite by accident. In early 1991, while visiting my family in North Carolina, I visited the state health department in downtown Raleigh to pick up some vital statistics reports, which I had begun to work with. One of these reports was on cancer deaths; and a simple eyeballing of the numbers in the report suggested to me that cancer deaths were rather high in some of the mountain counties in the western part of the state. That night, calculating some basic rates showed I had been right. Soon I was examining a map of the area when I noticed how close some of these counties were situated to Oak Ridge, Tennessee, site of the oldest nuclear complex in the world. Could nearly 50 years of emissions from Oak

Ridge have raised cancer rates among area residents? With that simple thought, I began to pursue the idea further.

After obtaining an enormous database from the National Cancer Institute on county-specific cancer deaths in the counties nearest Oak Ridge, I wrote a draft article. Since it was my first try at publishing in a medical journal, I (nervously) first submitted the draft to a number of eminent figures in the field of radiation studies. In addition to Gould and Sternglass, I solicited and received remarks from Samuel Epstein, a University of Chicago physician and researcher perhaps best known for his 1979 book *The Politics of Cancer*; John Gofman, a California-based health physicist who had become famous along with Sternglass beginning in the 1970s for his stinging charges of the harm low-level radiation was causing to society; Karl Z. Morgan, who had lived and worked at Oak Ridge for many years, and had begun the first national health physics society; and Alice Stewart, the British physician who started it all in the 1950s with her landmark study of prenatal pelvic X-rays and subsequent childhood cancer. The reviews all came in positive, and the article was published in mid-1994 by the *International Journal of Health Services*. The article showed:

1. Area-Wide Rate Up.   Between 1950–52 and 1987–89, the earliest and latest periods for which data are available, death rates for whites from all cancers combined *rose 31%* in the 94 counties within 100 miles of Oak Ridge, compared to an increase of only 5% for the U.S.

2. Rural Rates Up Faster.   The cancer death rate for the four urban counties in the area, containing cities like Chattanooga, Johnson City, and Knoxville, TN, plus Asheville, NC, went up 22.9%, while the rate for the other 90 counties, most of them rural or made up of small towns, had nearly double the increase (39.0%). Because the urban counties had already been exposed to more carcinogens, like air pollution from cars and factories, one would expect the rural rate, which previously were low, to climb faster. Sure enough, the urban and rural cancer rates were the same by the late 1980s.

3. Area Closest to Oak Ridge Harmed the Most.   The cancer death rate for Anderson County, Tennessee, where Oak Ridge is located, rose faster than in the 12 counties surrounding it (39.1% against 29.5%). This result also points to harm from reactor emissions, especially since Anderson County is made up of relatively well-to-do employees of the nuclear complex and their families, who have better housing, eat better diets, and have better access to medical care, and thus should be *less* affected by toxic substances.

4. Mountains Affected More than Lowlands.   Cancer death rates in mountainous counties near Oak Ridge rose faster than nearby lowland areas (40.4% to 30.3%). One would expect this to occur, as airborne radioactive emissions are brought down to earth by precipitation, and the mountain areas have among the largest rainfall totals in the nation.

5. Upwind Affected Less than Downwind.   Cancer deaths in the nearest three counties downwind (northeast) from Oak Ridge rose 50.8%, compared to only 7.1% in the closest upwind area. Prevailing winds blow more radioactivity to downwind areas over a period of time, so the results are consistent with the radiation-cancer connection.

Perhaps the most innovative point that the article made is that radioactive emissions from nuclear operations can affect an area as far as 100 miles away, not just the county closest to the reactor. The scientific community gave the piece only a passing glance. Now it was time to take the case to the public.

In July 1994, a press conference was scheduled at the Oak Ridge public library, sponsored by an antinuclear group made up of local residents afflicted with a variety of immune-related illnesses such as cancer. Because my health prevented me from attending, Jay Gould capably filled in for me. He returned to tell me that, compared to the largely apathetic or hostile scientific community, there was quite a stir among the public. At the entrance to the press conference, employees of the Oak Ridge nuclear plant were handing out leaflets titled "Mangano Study Discredited," and offering vague criticisms of the research. Other nuclear plant officials and state health department personnel were giving interviews to the press. Inside, Gould was faced with several sharp questions from officials and was interrupted when he tried to answer. In the weeks and months to come, however, I got to know a number of Oak Ridgers interested in the health issue. Conversations with them showed a split decision among the public. Some dismissed the study as of no consequence, but these tended to be Oak Ridge employees and their families, who depended on the nuclear plant for their livelihood. Many others, however, had caught the main point of my study: the Oak Ridge area, which once had a low cancer death rate, now exceeds the U.S. average, especially in areas that probably received the heaviest emissions from the plant's nuclear reactors since the 1940s.

In the years since, the health issue around Oak Ridge did not die. The state health department formed a committee that laboriously constructed an estimated dose received per person in the local area, a project that was nowhere near complete by early 1998. In 1995, the opening of the nearby Watts Bar nuclear reactor, which may be the last nuclear power plant to begin operations in the U.S. for a long time (none have been ordered since 1978 and none are being constructed), caused a stir. Protestors attempted to speak out at a public meeting, and some were roughed up and arrested by over-eager policemen, but the plant soon began operations nonetheless.

Concerns also grew about the Toxic Substances Control Act incinerator that started in 1991 at the Oak Ridge complex. The incinerator accepts low-level nuclear waste (e.g., clothing worn by workers) from plants around the nation, and burns it, emitting radioactive products into the atmosphere. In late 1997, I received a call from Bill Reid, an oncologist who became a cause célèbre in Oak Ridge by publicly stating several years ago that local residents had an exceptionally high rate of cancer, many of which grew aggressively.

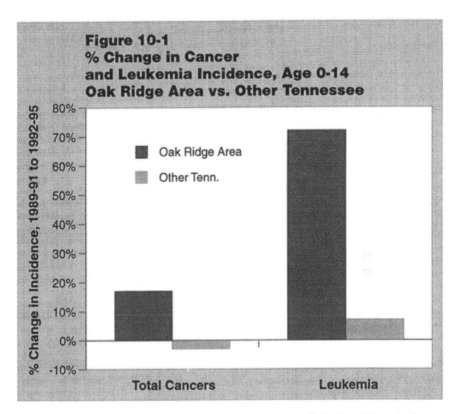

Figure 10-1
% Change in Cancer
and Leukemia Incidence, Age 0-14
Oak Ridge Area vs. Other Tennessee

Reid subsequently had trouble keeping privileges at the local hospital, was no longer referred patients by other physicians, and was not reimbursed by local private health insurance companies. Essentially blackballed from practicing medicine in Oak Ridge, Reid was forced to find work elsewhere, eventually taking a position in Nashville, but he remains interested in the Oak Ridge radiation health issue. He called to tell me that he and his wife Sandra, a nurse and antinuclear leader, had new concerns: "We've been getting some reports that there's a lot more childhood cancer, especially leukemia, in the area," he said, "and we think it's because of the incinerator." I contacted the Tennessee Cancer Registry and found that Reid had been exactly right. In the four years after the incinerator had opened, leukemia among children under 15 in the area within 50 miles of Oak Ridge had rocketed an unheard-of 72%, while that of the rest of the state rose only 7%. All childhood cancers combined had jumped 17%, compared to a decrease of 3% for other Tennessee counties (Figure 10.1).

So I learned that it is possible to convey the message to the general population that America has a public health problem, past and present, with radioactive emissions from weapons and power plants. But I also realized that a study like my Oak Ridge piece needed to be strengthened. Specifically, even though all acknowledge that the reactors emitted radioactive chemicals over the years, there is no precise calculation of how much was absorbed by the local population. Furthermore, looking at all cancers combined is important; but this statistic is very general and many factors can influence it. These vagaries tend to turn scientists off. I needed a more precise cause-and-effect demonstration addressing specific radiation doses and changes in specific diseases that scientists would recognize.

Disease-specific statistical studies

Chernobyl fallout in the U.S.

I got the precision I needed by turning my research to the effects of Chernobyl in the U.S. The low levels of radiation that reached America after the explosion were documented weekly in 60 American cities for three specific radionuclides by the Environmental Protection Agency (Chapter 4), so I already had a good "dose" part of the dose-response relationship. In addition, there were other scientists beginning to document rises in disease rates after Chernobyl. Beginning in the mid-1990s, a number of articles appeared on the enormous increase in thyroid cancer among children in Belarus and the Ukraine, which had received high doses of radioactivity, starting four years after the 1986 accident. Other articles began to show that young recipients of low-level emissions (Greece and the former West Germany) had higher risks of leukemia.

In a span of less than two years, I put together research showing unusual increases in rates of thyroid cancer, newborn hypothyroidism, infant leukemia, and childhood cancer in America after Chernobyl, specifically:

- Thyroid cancer rates in several U.S. states rose 19% in the early 1990s, compared to only 5% in the 1980s. The sudden jump in thyroid cancer cases *beginning in 1990* paralleled the change among children in the Ukraine and Belarus.
- Newborn hypothyroid cases in the U.S. rose 8% in 1986–87; like thyroid cancer, the increase was tied into larger amounts of thyroid-seeking iodine from Chernobyl in the food chain. The greater rate followed four years of no change. Moreover, the greatest rise occurred in the Pacific northwest (hardest hit by Chernobyl fallout) and the smallest change took place in the southeast (least affected by Chernobyl fallout).
- Leukemia cases before the age of one were 30% higher for U.S. children born in 1986 and 1987, or those who received exposure to Chernobyl fallout while still in the fetal stage, compared to those born in the other eight years of the decade. Humans are most susceptible to radiation before birth, and leukemia risk is greater after exposure to bone-seeking chemicals present in Chernobyl fallout like strontium and barium.
- Childhood cancer rates, after remaining largely unchanged from 1973 to 1982, rose an astonishing 37% from 1982 to 1993. The largest spike began with children born in 1986 and 1987.

I felt grateful that three of these analyses were published in well-respected peer-reviewed scientific journals, such as the *British Medical Journal*, *Lancet*, and the *European Journal of Cancer Prevention*. As of this writing, the article on childhood cancer has been submitted to the *International Journal of Health Services* and is under consideration for publication.

The press showed interest in some of the articles, and perhaps more important, I received dozens of letters from researchers and scientists across the world asking me for reprints. However, despite the success, I also encountered a major obstacle. With the exception of the *International Journal*, all of the above publications are British. My attempts to be published in U.S. journals met with complete rejection. The hypothyroid article, later printed in the *Lancet* (one of the preeminent medical periodicals in the world) was lambasted by the *American Journal of Public Health*. In a February 22, 1996, letter addressing my proposed article, the journal's reviewer raised a number of questions about whether the data I used were accurate and whether my analytical methods were appropriate. The letter's final point, however, gave away the (anonymous) reviewer's bias. "By far the most important consideration however is the biological implausibility of the hypothesis..." it read, meaning even if the data showed an iodine-hypothyroid connection, it couldn't possibly have been true. The article never had a fair chance.

The coup de grace was delivered in a cover letter expressing regret that my research could not be published. The letter was signed by Mervyn Susser of the Columbia University School of Public Health, who had done considerable work showing that emissions from Three Mile Island had caused no harm to human health, even going to the extreme of asserting that "stress" had caused any excesses in cancer and other diseases after the 1979 partial meltdown. A second blow came in the late summer of 1997. After submitting my original data on childhood cancer to *The Journal of Pediatrics*, I received a quick rejection letter from this eminent journal, which often carries articles on trends in childhood illnesses. The letter accused me of not documenting how my information was collected (when I clearly stated that all information came from state and local cancer registries). It also noted that my references of other evidence in the literature of rising childhood cancer rates were "poor examples of what data is currently available in the literature." In other words, new and unusual findings are not welcome.

Fermi

So while I had made headway in establishing myself in the medical literature and had drawn the interest of many scientists, there was still a large wall to climb among the pronuclear forces, especially in the U.S. So now, after having addressed the public (Oak Ridge article) and the professionals (Chernobyl articles), I moved on to an effort that would affect both of these audiences. Again, the evolution of this dual effort didn't come out of any grand design, but seemed to create itself as time went along. As I compiled my paper on childhood cancer, I would periodically call my colleague Ernest Sternglass and fill him in on some of the discoveries I was making. One of these new bits of information that really drew Ernest's attention was that although all areas with tumor registries showed a rise in childhood cancer incidence after the early 1980s, the area with the largest jump was the three Michigan counties (Macomb, Oakland, and Wayne) making up metropolitan Detroit. Ernest replied that Detroit was located only 30 miles downwind (northeast) from the Fermi 2 reactor. Fermi had had a long history of safety problems. The original Fermi reactor, for which planning had begun in 1951, had an accident on October 5, 1966, when it was still being tested. The reactor's cooling system failed and the core experienced extremely high temperatures. Even though a larger, Chernobyl-style accident was avoided, radioactivity did escape into the local environment. The original Fermi reactor never began operations and was dismantled in the mid-1970s.[1] In June 1985, a second Fermi reactor opened at the same site. In the late 1980s, the plant experienced many problems, among them fuel leakages, and a considerable amount of radioactivity was released into the atmosphere and the local water (the plant is located on the banks of Lake Erie). The Environmental Protection Agency's reports on radioactivity in pasteurized milk showed that in 1989 and 1990, levels of iodine-131 and cesium-137 in Detroit's pasteurized milk were well over comparable levels from 1983 to 1985, before Chernobyl. Subsequently, after a turbine explosion in late 1993, the plant shut down for needed repairs for over half of the period December 1993 to May 1997.

Ernest informed me that a number of concerned citizens were demanding an investigation into safety practices by the Fermi operators. A lawsuit by the local public advocacy group Citizens Resistance at Fermi Two (CRAFT) against Detroit Edison was about to unfold, he explained, and he had been called on to submit an affidavit at the trial. Information like rising cancer and leukemia rates among local children would be critical, he said. Would I be able to help him out?

Because of the existence of a local cancer registry, plus numerous other vital statistics, Ernest and I saw an opportunity to make perhaps the most comprehensive statistical demonstration of harm in any radiation-related lawsuit, including those brought by Nevada/Utah residents living downwind from the bomb test site, military personnel stationed near bomb tests, and residents in areas near nuclear reactors, including Three Mile Island. Within a few weeks, we had developed a health status profile of children in the Detroit area that had radiation-induced health crisis written all over it, including:

1. Childhood Cancer and Leukemia.   In the three-county area, cancers diagnosed before the age of five were 50% higher for children born in 1988–89, when emissions from Fermi were at their height, than for those born in 1980–85, before the plant started operations. For leukemia, the 1988–89 rate was 10% greater.

2. Infant Mortality.   In Wayne County, where Detroit is located, the death rate for white infants under one year fell an average of just 1% from 1985 to 1990, compared to an annual decrease of 3% in the post-

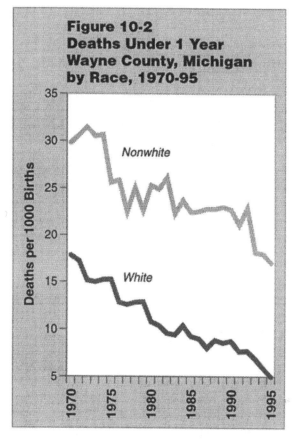

Figure 10-2
Deaths Under 1 Year
Wayne County, Michigan
by Race, 1970-95

bomb test era (1965 to 1985). For non-whites, many of them poor blacks living in Detroit's inner city, there was no change after 1985, after an annual 2% decline from 1965 to 1985 (Figure 10.2). Trends after 1985 looked very similar to those of the 1950 to 1965 period, when fallout levels in the diet were high due to the atmospheric bomb tests in Nevada.

3. Fetal Deaths. Wayne's record of deaths to fetuses after at least 20 weeks of gestation was also poor, after decades of decline. From 1985 to 1989, the fetal death rate fell just 6% for whites and surged 45% for non-whites.

4. Low-Weight Births. In Wayne County, from 1986 to 1991, the rate of babies born under 5½ pounds rose 8% for whites and 9% for non-whites.

5. Deaths from Birth Defects. From 1982 to 1985, deaths in the three-county area from causes related to congenital anomalies/birth defects averaged 206 per year. From 1988 to 1990, the average was 205 a year, unchanged despite the dropping rates all around the nation. The stagnant rate followed over two decades of steadily falling rates in the Detroit area.

6. Newborn Hypothyroidism.   The 1988–89 rate of congenital hypothyroidism in Michigan was 32.3 per 100,000 births, or 20% higher than
the national rate of 27.0. A majority of the state's births occur in the
lower part of Michigan, closest to the Fermi plant.

There are also indicators of immune disease proliferating after 1985
among the entire population, not just children in the Detroit area. Baby
Boomers appeared to take the worst of the health decline. Between 1983 and
1992, the total number of deaths age 35 to 44 in the Detroit area *jumped 24%
and 56%* for whites and blacks, respectively. One could argue that this
increase occurred due to a large population, more AIDS cases, and more
violent deaths. But this hypothesis falls flat on its face after considering that
deaths age 25 to 34 dropped 1% and 13% for whites and blacks, respectively,
over the same period. Persons in their late 20s and early 30s are also at
greater risk for deaths from AIDS and violent crime. So why the difference?
Once again, the year of birth stands out as the defining factor. In 1992,
persons age 35 to 44 were born from 1948 to 1957, meaning that their immune
systems were at maximum risk from bomb test fallout. In 1983, the age group
consisted of persons born from 1939 to 1947, conceived and born before
bomb testing moved to Nevada.

The number of Detroit-area deaths from 1983–85 to 1989–91 rose 30%
for pneumonia and influenza (Figure 10.3), 10% for breast cancer, and 28%
for asthma. It is clear that children and middle-aged adults suffered from
unexpectedly high rates of immune-related diseases and deaths after 1985,
and emissions from the Fermi reactor must be viewed as a potential contributor. As of this writing, the data are not being used in any legal action —
the CRAFT suit was dismissed, as were two appeals — but can be used in
future legal actions or public education efforts. They were presented in
March 1998 at a conference in Muenster, Germany, and hopefully in a future
article in a medical journal.

Because recent emissions of radioactivity were involved in the Detroit
case, the growing battery of public health statistics was available as evidence
in the case. Several decades ago, there was little more than infant mortality,
fetal deaths, and low-weight births to measure infant health. Now, especially
in an area like Detroit that operates a cancer registry, a wide range of health
status indicators can deepen the understanding of the "response" part of the
dose-response relationship between radiation exposure and health effects.

Clinical studies

There is yet one more step needed to convince scientists and the public —
some of whom are already convinced — of the toxic effects of a substance
like radioactivity. Clinical studies, in which measurements of actual *in vivo*
radiation are taken, or estimates of how much radioactivity was ingested,
are the final frontier. The history of collecting these types of data is a poor
one. Again, taking health effects from Hiroshima and Nagasaki survivors

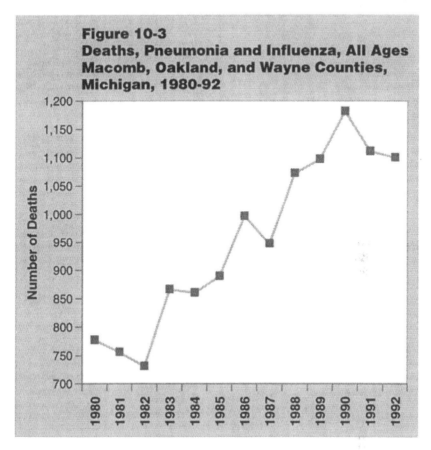

Figure 10-3
Deaths, Pneumonia and Influenza, All Ages
Macomb, Oakland, and Wayne Counties,
Michigan, 1980-92

and assuming the same *per-dose* effects occur with low levels of radiation slammed the door on low-level radiation studies by government health and safety officials. Even though Petkau's theory changed some minds, statistical data indicting routine emissions from nuclear reactors proliferated, and many citizens grew concerned about the risks of living near nuclear plants, officials were not about to ante up funds for these studies. Plant operators have monitored radiation doses absorbed by workers at nuclear plants over the years, but even a series of analyses revealing higher risks for cancer and other diseases have not prompted the nuclear industry or the government (public health officials and the military) to carefully check effects on the nearby population. Any studies done on animals used high doses of radiation: if there were any analyses using low-level, long-term radiation exposure by industry or government-funded researchers, none were ever made public.

The only truly clinical studies that have been done are few and far between. As mentioned in Chapter 3, a St. Louis advocacy group began collecting baby teeth in 1958 from local six- and seven-year-old children and analyzing them for strontium-90 content, since this chemical lodges in bone and bone-like substances such as teeth. The startling finding that the strontium-90 concentration had risen 30 to 40 times in children born in 1965 (peak concentrations of strontium-90), compared to those born in 1951 (the start of the Nevada above-ground bomb testing) played a role in educating health professionals and the public of the dangers of bomb test fallout, leading to the Partial Test Ban Treaty of 1963. After that, either nobody thought of replicating the St. Louis study (or performing a new study to examine the amount of radiation absorbed from nuclear operations) until the mid-1990s. Even then, it took the catastrophic experience of the accident at Chernobyl to begin such a probe. As stated in Chapter 4, it was German researchers who tested baby teeth for strontium-90, finding that German children born in 1987 (the aftermath of Chernobyl) had nearly 10 times the concentration in their baby teeth than did those born between 1983 and 1985 (before Chernobyl).

A recent effort in the United States will also assess strontium-90 in baby teeth, but instead of comparing persons born before or after a nuclear event like bomb testing or Chernobyl, it will compare those living close to and far from a nuclear installation. The idea, first proposed by Jay Gould, will compare strontium-90 in the teeth of children living near the safety violation-plagued Brookhaven research reactors on Long Island, and compare these findings with children living at least 50 miles from Brookhaven or any other reactor, in the New York City boroughs of Brooklyn and Queens. Gould and others are trying to get word of the study to interested parties in other areas, since the methodology can be applied to any of the 60-plus areas near a U.S. power or weapons plant. No results were available as of early 1998, as the process of collecting an adequate number of teeth for analysis continues, but when results are in and are compared with rates of cancer, infant deaths, and underweight births, close to and far from Brookhaven, a new understanding of effects of low-level radiation may ensue.

A final clinical study measuring the actual dose of radioactivity absorbed by civilians was the October 1997 National Cancer Institute estimate for each county in the U.S. of the iodine-131 absorbed during the 1950s. A simple visit to the world wide web can provide information on how much iodine-131 was absorbed by the average individual in a particular county, according to when he/she was born. Nobody has yet had the chance to correlate this enormous amount of data with actual health effects, but the database now exists to do this. Drawbacks include the lack of a cancer registry in many states and the fact that only one element (iodine-131) of the atomic cocktail is covered, but the NCI has certainly provided a starting point for future research.

Another important way to connect radiation with any resulting health effects is through a *case-control study,* which has yet to be done. In a case-control approach, the researcher tests a number of persons with a disease (children with leukemia, for example) for the level of a suspected contributor to the disease (strontium, for example). The sick persons are known as the "cases." In addition, the researcher selects a similar number of healthy persons of the same age, race, and gender composition as the sick population, along with other similar factors (e.g., living in the same area, having the same socioeconomic status, etc.) and tests them for the same substance, such as strontium. The healthy persons are known as the "controls." If the strontium content of the cases is significantly greater than in the controls, there is a clinical correlation between the offending substance and the disease. If strontium content is the same, or is less prevalent in the sick persons, there is no correlation, and the theory that the toxin caused a harmful effect is discarded. Such an effort for the effects of low-level radiation has never been done and will probably not be done in the foreseeable future, but would elevate the level of understanding of the effects of low-level radiation immensely.

Statistics ride the PR train — Brookhaven

The actual and hypothetical analyses described above have all relied heavily on statistical and/or clinical data. What's largely missing in these, however, is publicizing the information through the democratic process of mobilizing popular support which in turn galvanizes leaders to make necessary changes. For years, studies had shown that racially segregated, separate but equal classrooms were in fact not equal, and bred a second-class self-perception in minorities; but it took a Supreme Court decision and years of civil rights activities to finally erase legally sanctioned Jim Crow schools. The environmental movement of the late 1960s and early 1970s took the growing body of evidence that certain chemicals were harming the population and engaged in highly public demonstrations, rousing state and federal governments to stop producing the offending substances.

One of the current drives to identify and eliminate harmful low-level radiation is taking place at the Brookhaven site on Long Island, New York. Efforts against Brookhaven are different than those at most other nuclear sites, however, in that they are strongly driven by effective populist movements and publicity. Brookhaven National Laboratory, located in eastern Long Island, was one of the plants designed after World War II primarily to lead the research effort to strengthen the country's nuclear weapons program. Opened in 1950, it has operated three relatively small reactors, and has succeeded in attracting eminent scientists, including four Nobel Prize winners, to its staff. For years, Brookhaven Associated Universities, nine Ivy League-caliber institutions, have run the plant.

Maybe the lack of safety knowledge, the urgency of expanding the nuclear weapons arsenal, or sheer arrogance by operators led to steady releases of radioactivity beginning in the early 1950s. In 1972, Sternglass presented data showing liquid waste emissions from Brookhaven grew steadily until it reached a peak in 1961. He correlated this trend with a 1955–61 Suffolk County infant mortality rate 7% greater than in 1949–50, before the plant began operations.[2] Beyond this paper, however, no known research on health effects from Brookhaven emissions was performed until the 1990s. Shabby health and safety practices continued for years and were ignored by plant operators and regulators. Even when the New York State

Department of Health found radioactive tritium in amounts 100 times normal in groundwater beneath the reactor complex in 1992, plant operators essentially did nothing to monitor or clean up the problem.

The 40-plus year history of contamination and arrogant denial about any potential public health problem came to an end beginning in 1993. Actually, groundwork for the public exposé of unsavory Brookhaven practices had been laid long ago. Breast cancer rates for women in Long Island had been growing in leaps and bounds, causing much unrest among local residents. The politicians and media picked up on the story, but fumbled around to identify a cause(s) for this epidemic. Public health agencies were stonewalling any investigation into potential environmental factors. In 1993, the Long Island breast cancer coalition One in Nine was told by the National Cancer Institute that the problem had no environmental cause, but was attributable to the presence of large numbers of Jewish women.[3]

At this point, the publicity machine that later fingered Brookhaven as a suspect kicked into gear. The West Islip, Long Island One in Nine chapter responded to the outrageous NCI stance by conducting their own door-to-door survey of breast cancer in the town, making a large map of cases and deaths and presenting it at public hearings. The West Islip group got a large amount of media attention that provoked public anger, and public health agencies suddenly changed their policies. In 1994, NCI began a five-year study on causes of breast cancer on Long Island. For the first time, the state health department released town-by-town rates of the disease in the 1980s. Astonishingly, the highest rates in Suffolk County, and among the highest in all of the U.S., were found *in the 24 towns situated within 15 miles of Brookhaven.*

The cat was out of the bag and a panicky Brookhaven circled its wagons and tried to discredit its opponents' case or made moves to appease the population. Brookhaven's leaders knew what they were facing: in a post-Cold War world no longer building or testing atomic weapons, its existence was not as crucial as it once had been. Suddenly on the defensive, Brookhaven fielded shot after shot, not in the quiet of medical journals or professional meetings, but in full view of the public:

- Late in 1995, Brookhaven representatives declined to meet Gould and Long Island environmental activist William Smith face-to-face on public access television. Instead, they presented a critique devoted to discrediting the methods showing high breast cancer rates near the plant. The result was a flood of anti-Brookhaven letters to eastern Long Island newspapers, and several town meetings held on the subject.
- In late 1995, Brookhaven operators planned to upgrade its sewage system by dumping 1 million gallons of contaminated water into the nearby Peconic River. Smith's group, called Fish Unlimited, began to prepare legal action to prohibit the dumping; and after being pres-

sured by local politicians and businessmen, Brookhaven reconsidered its plan.

- In January 1996, a week-long series by WPIX (a major independent New York City television station) based on information supplied by a whistle-blowing lab employee documented a series of Brookhaven safety violations and coverups. The series included an unsettling videotape of radioactive waste being discharged into groundwater. Shortly after, New York Senator Alfonse D'Amato, a Long Island resident, called for Senate hearings into practices at Brookhaven.
- In late 1995, Brookhaven tried to soothe public concerns by offering free bottled water to residents just south of the plant using private wells. After years of denials of a public health threat, BNL now acknowledged that groundwater was contaminated as deep as 200 feet below the surface, and the pollution was moving south. This gesture only appeared to make local residents more irate; a fiery meeting in the nearby town of Shirley also made the January 1996 news program.
- In April 1997, the NBC-TV affiliate in New York City telecast a story of children living near Brookhaven stricken with rhabdomyosarcoma, a rare soft tissue tumor. Eight such children had been identified, over 100 times the number normally expected in a population that small.
- In May 1997, after 500 Long Island women sent a petition to Senator D'Amato calling for action, the U.S. Department of Energy fired the research team from the nine prestigious universities that run the plant. Six months later, DOE replaced them with a group from the state university at Stony Brook, Long Island. While the firing was an unprecedented DOE action, it was still met by skepticism. Local Congressman Michael Forbes welcomed the change, but challenged the Stony Brook team to "expedite the cleanup of the entire facility."[4]
- In November 1997, the U.S. General Accounting Office issued a report on Brookhaven's long-standing tritium leak from its aquifer and its laxness in monitoring and correcting the leak. The report was a large blow against the Brookhaven operators and the DOE.
- In January 1998, *The Montel Williams Show*, seen nationwide by millions, devoted a full hour to the problems at Brookhaven. Actor, activist, and local resident Alec Baldwin, whose mother suffers from breast cancer, spoke passionately about the contamination and the need for action. Physician and nuclear expert Helen Caldicott and attorney/activist Jan Schlictman also appeared. Once again, Brookhaven officials declined to appear on camera, but submitted a videotape assuring that the plant's reactors were operating safely.

By early 1998, Brookhaven was still in operation, but very much under the gun from politicians, activists, and the public at large. Many Long Islanders, both lay public and professionals, were convinced that Brookhaven had contaminated the local environment, which contributed to higher rates of breast cancer. However, unlike Oak Ridge or Fermi, the process leading to this new understanding had largely been fueled by *public* manifestos, such as town meetings, media coverage, and grassroots co-opting of politicians and businessmen. Articles and research played a part in these public actions, but on their own, they would have been destined to sit for years gathering dust on a library shelf.

Summary

In summary, proving (or convincing the majority of professionals and citizens) that a substance is hazardous is a rigorous process. In the case of low-level radiation, which has been at the center of a hot political controversy involving the tradeoff between national security and the public's health for decades, this process is especially tough. Dollars for objective research are hard to come by, because the purse strings are held by officials who are not interested in exploring a topic they think is bunk; or, if proven true, may cost the government and the nuclear industry dearly in terms of restitution to victims and public esteem. However, as casualties mount, things are changing. Recent revelations that the government and industry were less than honest with the people about radiation's effects also sways the public opinion. Just in the last four years alone, the federal government has admitted to conducting over 200 atomic bomb tests that had been kept secret for decades, experimenting with dangerous radioactive products on patients without obtaining their consent, explaining that above-ground bomb testing could have caused up to 75,000 additional thyroid cancers than previously thought, releasing health records of nuclear workers for the first time, and closing down a number of nuclear reactors for health and safety reasons. There is a long way to go, but this is a start. Enough such discoveries and revelations, especially in the post-Cold War era when the nuclear arms race is no longer an issue, will drive public opinion to the point that elected and regulatory officials have no choice but to allow objective research to finally be conducted, and appropriate changes in nuclear policy to be made. This research will help solve the immune mystery surrounding Baby Boomers and their post-1983 descendants, and prevent some of the past changes from repeating themselves in the future.

References

1. Fuller, J., *We Almost Lost Detroit*, Reader's Digest Press, New York, 1975.
2. Sternglass, E., Environmental radiation and human health, in *Proceedings of the Sixth Berkeley Symposium on Mathematical Statistics and Probability*, University of California Press, Berkeley, 1972, 191, 207.
3. Gould, J., *The Enemy Within*, Four Walls Eight Windows, New York, 1996, 150.
4. *The New York Times*, November 26, 1997, B5.

chapter eleven

Radiation exposure and the future: great fear and great hope

The information provided in this book is disturbing. It has revealed massive immune health deficits among Americans born from 1945 to 1965 and after 1983, who now make up about half of the U.S. population and represent the leaders of the present and future. And the book didn't touch effects of radiation on other Americans. With disease and death rates rising, and millions possibly suffering subtle effects of exposure, is there any hope for the future, or have the misjudgments of the past half-century permanently crippled our society's health, creative abilities, and production?

Because the future can't be predicted with certainty, the answer in 1998 remains unknown. But I can foresee many things happening. These predictions include a worst-case and best-case scenario, or a mixture of the two. Following is a description of both extremes. Each is based on a set of human decisions and actions which affect society's health and welfare, as actions have done in the past 50 years (Chapters 2 through 4).

Worst-case scenario

Whenever nuclear power is discussed, naturally the greatest fear is any use of atomic weapons in warfare. Despite the fact that none have been used since 1945, that tensions of the Cold War are over, and that treaties beginning in the late 1980s are drastically reducing the world's nuclear stockpile, the threat of nuclear war persists. The U.S. and former Soviet Union still have enough warheads to end life on the planet as we know it, and a half-dozen more nations have nuclear weapons capable of wreaking horrors far greater than those at Hiroshima and Nagasaki. Yet, until the world rids itself of every last atomic weapon, or finds a way to negate a bomb's effects, this threat will continue to be a fact of life.

But in addition to the possibility of nuclear war, there are a number of actions or inactions related to atomic energy that may lead the U.S. down a path of sickness, death, and societal decline. Each of these is described below.

Resumed bomb production and testing

In the late 1980s, as the Soviet empire unraveled, the U.S. stopped producing fuel for nuclear weapons, and no new warheads have been produced since. On September 23, 1992, a weapons test took place in a shaft deep under the Nevada surface. Since the detonation of this bomb, with the approximate power of the weapons used on Hiroshima and Nagasaki, no tests have occurred in the U.S. With the exception of the 1996 French underwater test of a bomb in Muroroa in the South Pacific and the 1998 underground tests of five weapons by India and six by Pakistan just weeks later, bomb tests have ceased throughout the world during the past six years. Discussions are under way for a permanent Test Ban Treaty. Even though this development is a welcome one, it will be a challenge for the world's political leaders to continue the ban. It only takes a single leader's decision to threaten the ban, such as the Iraqi dictator Saddam Hussein's decision to pursue bomb development just before the Persian Gulf War. Any restructuring of the world order could create an imbalance of power motivating one nation to produce and test new bombs, and other nations to match it. In addition, some American military leaders are pressing to resume testing for "quality assurance" purposes, a contention that some dispute. We have history as a guide that this could happen: in 1961, after years of harmful above-ground testing, the U.S. and U.S.S.R. broke a three-year moratorium and showered the planet with the most massive amounts of radioactivity, until the near-disastrous Cuban Missile Crisis brought all to their senses. Even if tests are conducted underground, and if weapons production is "cleaner" than it was in the middle of the century, the addition of low-level radiation to the environment would be a harrowing re-visit to the past, with potentially dire health consequences.

Continued reactor operation

At present, there are 109 nuclear reactors operating in the U.S., producing about 20% of the nation's electricity. Both these figures are at all-time highs, although they fall well short of President Nixon's 1971 prediction of 1000 reactors in the country by the year 2000. The figure of 109 should not go any higher, since no new nuclear reactors are being planned, and the last order for a reactor was placed in 1978. After the percentage of U.S. electricity produced by nuclear power jumped from 11 to 20 from 1980 to 1988, the proportion was still stuck on 20 in 1997. "We don't have gains for nuclear" in future projections, says DOE's Jay Hakes.[1]

Many reactors have histories of safety and health problems described in this book and elsewhere, in an industry where there is simply no margin for error. U.S. reactors operate at only about 60% of maximum capacity, which contributes to a higher-than-expected cost to produce electricity. Thus, the dream that many had of nuclear power plants providing energy in a cheap, clean manner has never materialized. A number of the reactors are suffering from the wear and tear of extended use; the suit filed against Westinghouse by 14 utilities operating 15-year-old reactors over the defective and corroded steam generator tubes expected to last 40 years is an example of the environmental problems caused by age and faulty reactor design. Even the least harmful reactors opened in the 1970s are reaching the end of their useful life; in the first years of the 21st century, utilities running dozens of reactors will have to decide whether to renovate or abandon them. This decision will not be made just by the utilities, but must be approved by the national and state regulatory agencies responsible for maintaining industrial safety and protecting the public's health. Historically, economics has taken precedence over safety and health in planning the operation of nuclear power plants, and if these past practices do not change, and hazardous nuclear plants are not terminated, an extension or exacerbation of the immune disease epidemics plaguing America will likely occur.

Cutting corners

Although no nuclear plants have been ordered in this country for 20 years, there is no guarantee that this will continue. History once again offers precedents that public policy can encourage more nuclear power generation. In the 1950s, Congress passed the Price–Anderson Act limiting liability of nuclear plant operators, and the once-jittery utility companies began to order large numbers of reactors. In the early 1980s, the Reagan administration tried mightily to reduce the average time needed to open a plant from 14 to 7 years, by "cutting excessive red tape," which was mainly standards to protect the public's safety and health. With America's future energy policy very much up in the air, and if another energy crisis raises costs and rouses the public into calling for a quick source of cheaper energy, the pronuclear forces could well persuade anxious public officials in make it easier to build and operate nuclear power reactors.

Downplaying waste maintenance and cleanup

Future policies on how to control and clean up waste are crucial to how future generations' health will be affected by radiation. Until now, no plan for permanent waste storage has ever been implemented by the U.S. government, but much of the high-level waste (i.e., spent fuel) has been stored at the respective sites of production. Even when these storage facilities leak, the waste has often been contained in the ground and groundwater. Thus far, local residents haven't actually ingested much of these poisonous chemicals, but to ignore the problem of cleanup will have much more serious consequences. In Chapter 4, we have covered the situation at Hanford, in which the leakage of liquid radiation is moving downward in the ground, and in several decades will reach underground water that feeds the Columbia River if its course is not reversed. Contaminated groundwater under Brookhaven has polluted private wells south of the plant.

If two steps aren't taken, the waste problem will manifest itself into the further erosion of human health. First, public support of nuclear waste clean-up now in progress must continue. This support is not assured because, since the waste poses no immediate danger, public officials often make programs like this a target for budget cuts. Private groups such as the utilities have shown little commitment to waste removal unless threatened by embarrassing public exposés and government mandates. Second, a decision must be made on a permanent plan for waste storage. For decades, this has been considered, and the government wants to make an underground site below Yucca Flats, Nevada, such a permanent home. But some have raised safety questions about the selection of this spot, like what would happen in case of an earthquake? The choice of a site must be risk-free for several thousand years, the time it takes some of the long-lived radioactive products like plutonium-239 to decay, or again, human health will be impaired.

Not regulating reactor sales

In the past, reactor manufacturers like Westinghouse and General Electric have been able to produce and sell machines that do not live up to their purported ability to maintain a safe environment. One such example of this practice is the assurance given by Westinghouse that its steam generator tubes would last the entire expected life (40 years) of a reactor, only to find these tubes were corroding and allowing radioactive leakages after about 15 years. With the U.S. demand for new reactors virtually dried up, manufacturers have turned their attention to foreign markets, especially third-world nations that have very underdeveloped systems of electrical power production, and equally underdeveloped safety standards. The Clinton administration's announcement in late 1997 that the U.S. would sell nuclear reactors to China is an example of American manufacturers and government sending reactors to overeager nations that have no ability or proven record of safely operating nuclear plants. Safety standards are set by sovereign governments, not the U.S., and monitoring must be done by the purchasing nation. Thus, a sale of reactors to a nation like China not only creates a risk of greater routine emissions of fission products affecting local residents, but of a catastrophic accident the magnitude of Chernobyl. A common misperception of the 1986 explosion was that it was caused by a faulty Soviet-made reactor. While there were flaws in the design of these reactors, the one and only reason for the Chernobyl accident remains *human error*. If an accident affecting the entire world occurred in the former Soviet Union, which at least had experience in atomic weapons and power production, we must fear the possibility of a repeat disaster in nations with less nuclear know-how than the U.S.S.R.

Lax government setting, monitoring of reactor operating standards

For decades, the issue of how strictly government regulators set and monitor standards for nuclear reactor operations has raised major safety and health concerns among Americans. The very design of the old Atomic Energy Commission as both a producer and promoter of nuclear energy made an objective appraisal of safety impossible. The AEC was terminated in 1975, and the Nuclear Regulatory Commission took over the role of safety enforcer, while public health agencies like the National Cancer Institute were responsible for evaluating and monitoring health effects of reactor emissions, as it attempted in its 1990 study for Senator Edward Kennedy.

But this setup hasn't worked very well either. For years, Northeast Utilities, which ran all four reactors in Connecticut (at Millstone and Haddam Neck) were never penalized for their substandard and sometimes dangerous practices. It took several whistle blowers providing information to *Time* magazine for a cover story in March 1996 for the regulators to take action. Almost immediately, the NRC shut all the reactors, and as of January 1998 none of them have been reopened because the NRC is not satisfied with the changes and plans made by Northeast Utilities. This series of events demonstrates how severe the problems were all along, without the NRC lifting a finger. It is insufficient to rely on a public scandal to erupt in the press before the NRC takes any significant action to curtail or close a reactor. The only way the public will receive adequate protection is for regulators to enforce the law to the fullest; but thus far, in the 50-plus years nuclear plants have operated, this level of vigilance has never been approached. Unless there is a major turnaround in the NRC and state radiation safety regulators, the public's health will suffer from this lack of diligence.

Continued denial by health officials

Similar to the poor performance by watchdogs of radiation safety standards, the history of America's public health officials insisting that low-level radiation is not harming workers or the public is a shameful one. Many officials are in complete denial that any low-level exposure can ever be harmful, and often refuse to address or purposely misinterpret information that suggests otherwise. As evidence, the National Cancer Institute never did any systematic study of cancer trends near nuclear plants; when finally forced to do so by Senator Kennedy in 1988, NCI researchers used a flawed methodology and outright denial of results to conclude that no pattern of elevated cancer rates exists near nuclear operations. This book has often cited the NCI study to indicate some of the adverse cancer trends in populations living near nuclear plants. Government has consistently given no support for any research that might indict extended low-level exposures in any way. Moreover, there is a litany of harsh reprisals for anyone who would dare question the idea that prolonged low-level exposures are anything but safe. The mid-1950s report of an AEC veterinarian studying thyroid damage in Utah sheep that had died in the aftermath of a Nevada bomb test was suppressed by the government for years, while the public was falsely told that the sheep had died of malnutrition. Pittsburgh researcher Thomas Mancuso was forced to retire after finding higher-than-expected cancers among workers at the Hanford site. Esteemed physicist and physician John Gofman was stripped of most of his staff and funds by the federal government after stating that atmospheric bomb tests had caused the deaths of several thousand babies. Even recently, the government took an astounding 15 years to complete a study showing that iodine exposures from bomb testing may have caused as many as 75,000 excess thyroid cancers.

The way to erase this sordid past is to adequately fund objective research on health effects of low-level radiation and make appropriate recommendations to public health officials and nuclear plant operators. But not a penny appears to be forthcoming from Washington or the private sector, and even independent research showing a low-level radiation/health effects linkage is suppressed (refer to my experience in submitting a letter to the *American Journal of Public Health*, explained in Chapter 10). Continuation of these attitudes and practices translates to additional Americans who will sicken and die.

What if this worst-case scenario comes to pass, and all these possible trends and events become reality? In terms of what happens to reactor operations, the potential situation is not a pretty one. The longer reactors operate, the more there are in operation, and the less compliant reactors are with safety standards makes several consequences more likely. The chance increases of another dreaded accident (domestic or foreign), contaminating the food and drink supply for months and years (á la Chernobyl). Continued emissions of low levels of radioactivity will cause more nuclear workers and civilians to ingest these toxic products. The inability to solve the nuclear waste dilemma brings us closer to a more widespread contamination of our water and food. And the continued supply of bomb-grade nuclear fuel extends the possibility of an accidental explosion or theft by a terrorist group.

In terms of human health, it is easy to see that this worst-case forecast would further extend the trends outlined in this book. The cancer monster would proliferate, along with other immune diseases. Chances are that medical science will continue to be more proficient at curing these diseases or at least extending life, but this is no solution. The costs of treatment to keep these people alive will become more staggering than it is today. In the case of children and young adults, victims of serious disease may survive, but will be taking medical resources and disability checks, rather than contributing manpower, ideas, and tax dollars to society. In the case of the millions of other Americans who are not dead or very sick, but have incurred subtle effects to their mental and physical development, the U.S. may find itself with a thinning of the "best and brightest" sector, and economically may fall even further behind nations like Germany and Japan. Finally, because blacks often seem to suffer more than whites from the effects of toxic substances such as radiation, the current racial divide may widen even further. This worst-case scenario will not wait long to play itself out: the Baby Boom generation is already moving into the age groups in which cancer and other immune diseases reach their highest rates. A health disaster is looming.

Best-case scenario

While the previous few pages may be depressing and gloomy, there is another path that can just as easily be followed. Each of the above possibilities would raise the level of environmental radiation and increase risk for higher rates of numerous diseases. Presented below is the reverse; in other words, alternate policy decisions for each issue addressed would minimize the amount of radiation exposure for the American public, and lower the risk of suffering and/or dying from certain diseases. I have not compiled a "pie in the sky" set of events in which all pronuclear leaders would do an immediate and complete about-face and make the world completely free of the danger of low-level radiation. Rather, I have suggested paths that are quite realistic and possible in the near future.

No further weapons production and tests

The U.S. and all other nuclear nations could continue to adhere to the voluntary no-test policy begun in late 1992. Perhaps testing would be banned or severely restricted altogether in a comprehensive and mutually verifiable Test Ban Treaty now under discussion. If this scenario became a reality in the early 21st century, the health of American (and world) citizens would benefit greatly. If weapons production ceased, the old and dirty sites run by the U.S. Department of Energy would focus on tasks like medical research, evaluation of the damage caused by production, and cleanup of used fuel and other nuclear waste, instead of continued human exposure from fission-producing operations. Cessation of bomb tests themselves would lower health risk by eliminating the chance of an accident, such as Nevada's Baneberry (1970), Misty Rain (1985), and Mighty Oak (1986) tests that leaked radioactivity into the environment. It is important to know that this goal could be reached without the sacrifice of any defensive military strength. Nuclear powers like the U.S. and Russia would still have plenty of weaponry to completely counter any nuclear attack.

Maintaining/Enforcing strict safety standards

Despite the disappointing performance of regulators like the AEC and NRC to set and enforce meaningful policies of safety in nuclear power reactor operations, the possibility exists for future regulatory activities to truly protect the public. If this were to happen, health benefits would be extensive. Regulators would review current policies and revise them, especially in light of health effects documented in this book and elsewhere. Inspections would be frequent and impartial, with maintenance of public health rather than continued operation of nuclear reactors as regulators' top goal. Penalties would be swift, appropriate, and proactive, not simply a matter of reacting to bad press. Regulators should fine, temporarily shut down, or permanently close any reactor that does not meet strict safety standards regardless of economic or publicity consequences.

If this were to happen, a number of the 109 currently operating reactors would be shut down immediately or retired in just a few years. Would the NRC and state radiation safety officials completely reverse their field and make this happen? It's happened before with other health threats: regulators took on the powerful tobacco industry starting in the mid-1960s and sparked a 40% reduction in the number of cigarette smokers in the next 25 years through public education, advertising bans, funding of objective health studies, taxes, and other means. The nuclear industry — the utilities running plants and manufacturers of reactors — are powerful corporations, but they are not infallible. For regulations to stand up, they must be supported by elected officials, which really means the U.S. President. A strong-willed President would take on the nuclear industry and its agenda, much like Theodore Roosevelt knocked the mighty trusts on their heels in the early 20th century. The President would appoint officials to the NRC, Environmental Protection Agency, National Cancer Institute, etc., who are committed to establishing the truth and protecting the population, not just protecting an industry's profits. And to get a President like this, there must be a popular will demanding such actions. This is still very much a democratic nation, and if the people want something strongly enough, the politicians will respond, and enforce health-enhancing policies in order to remain in office.

Sharply curtailing foreign sales of reactors

Manufacturers of nuclear reactors like Westinghouse and General Electric are chomping at the bit to expand their business to foreign clients, such as the former communist nations of the Warsaw Pact and third-world countries. And although these companies would welcome a renaissance of American nuclear reactors, this market has been dormant for 20 years, and shows no signs of re-emerging. Selling reactors to foreign nations presents a dilemma, namely, that the purchasing nations have little or no experience in operating reactors, few meaningful safety standards, and no regulatory structure to enforce these standards. With the possibility of routine emissions, small-scale accidents, or catastrophic accidents like Chernobyl at stake, the U.S. government should *strictly regulate* any sales of reactors to foreign nations. All sales should be banned, unless the purchasing nation shows convincing proof that it has the knowledgable manpower to run reactors safely and the regulatory apparatus in place to ensure a high level of safety. Perhaps few nations could meet these criteria today, but selling no reactors abroad would benefit the health of the U.S. (and world) population.

Fund objective health research, and follow up on findings

With the political pressures of the Cold War well in the past, the American public health research community has a clear chance to tell the story of what radiation has done and is doing to society's health. Changes need to be made in all levels in research, which are identified in Chapter 10. The statistical or epidemiological study function must first receive more funding, from both public and private sources. Findings and interpretations of the research must be reviewed by a nonpartisan, objective panel of experts. If there is consistent disagreement between the researchers and reviewers, the researchers should be prohibited from receiving additional funding. Clinical research should proceed in the same way as the epidemiology component, with objectivity and oversight the keys. Medical studies should take place in special clinics, located near nuclear plants, dedicated to the understanding of how much and what types of radioactivity Americans have absorbed, and relating this information to clinical indicators (like white blood counts, thyroid hormone levels, etc.) as well as disease rates. All findings should be made public and used in future policy decisions governing nuclear operations. This new approach to research on radiation's health effects in America should be linked with similar efforts overseas to the greatest degree possible.

Public education program

Just as the government and nuclear industry made a concerted effort in the 1950s to show how nuclear energy could improve the lives of Americans, a similar commitment from the public health community is needed, focusing on the adverse health effects of chronic low-level exposures from bomb test fallout and from nuclear plant emissions along with the helpful uses of man-made radioactivity, such as in medical treatment. This kind of public education effort warning against a hazardous product has been done successfully before: beginning in the mid-1960s, health departments and other officials relentlessly fought to make people understand the lethal effects of tobacco use. Television and radio advertising was banned by Congress, increasingly strong warnings were placed on the labels of cigarette packages, curricula were introduced into high schools, and public service announcements were placed on the airwaves. All of these contributed to the massive drop in cigarette smoking. Another example is the new awareness of household radon levels, so that home-dwellers can understand the health implications of this radioactive chemical, check radon levels in their household, and reduce these levels as necessary. Groups at greatest risk for immune diseases caused by radiation, such as the Baby Boomers, should receive encouragement to have their immune system checked and to be screened for certain diseases. This idea is not a far-fetched one: Senator Tom Harkin, after learning about the National Cancer Institute's 1997 study indicating that iodine exposures in the 1950s caused thousands more thyroid cancers than previously thought, suggested that all Americans in their 40s and 50s receive a free thyroid exam.[2] These actions would return the public health community to its rightful role, that of an advocate for the public's health, and would offset its history of remaining silent or declaring that low-level radiation is harmless despite evidence to the contrary.

Evidence of hope

These, then, are the steps that can be taken to reduce future low-level exposures and reduce the sick lists and death tolls. But even with this cheerier promise of less low-level radiation and less disease, some may worry that we are doomed in the immediate future; the disease and death toll will still be enormous even if the safer path is chosen and radioactivity has already been ingested and its ill effects cannot be reversed. Certainly, we are dependent to a degree on our leaders, public and private, to make the changes needed to reduce exposure and improve health, but it is critical to know that individuals can do more than sit, hope, and pray. There are some basic steps that prevent additional exposures to radiation, such as not living near or working at a nuclear facility. But members of society can be even more proactive to *improve*, not just protect, their own health.

Two articles appeared in the medical literature in 1997 showing how certain adverse immune-related outcomes in newborns could be lowered by mothers taking modest doses of vitamins during pregnancy. The first covered over 1100 poor women, just over half of them black, in Camden, New Jersey. The pregnant women who took a basic multiple vitamin every day starting in the first trimester of pregnancy reported preterm (under 37 weeks gestation) deliveries 47% *less* than those mothers who took no vitamin supplements during pregnancy. The rate of very preterm (under 33 weeks gestation) deliveries *dropped an amazing 78%.* The risk of low birth weight, which has been repeatedly emphasized in this book as being sensitive to radiation, dropped by roughly the same amounts, *even when the mother began taking vitamins in the second trimester of pregnancy.*[3]

The second article, involving 323 pregnant women from Dublin, Ireland, looked at the percent of babies born with neural-tube defects. These types of abnormalities, for which risk is known to be raised after exposure to radiation, include conditions such as spina bifida. The study gave women small but varying amounts of folic acid, which produces red-cell folate known to reduce risk of neural-tube defects. Results show that the risk of such a birth defect *drops 47%* among women taking 400 micrograms of folic acid daily during pregnancy (the recommended daily requirement set by the U.S. Food and Drug Administration) compared to women taking no folic acid supplement. Even a half dose of folic acid each day reduces the risk of neural-tube defects by 41%.[4] To ensure healthier newborns[5] the U.S. Food and Drug Administration is now attempting to ensure that folic acid-rich servings of grain products like rice, pasta, and bread contain at least 10% of a woman's daily requirement.

The findings in these two studies have enormous implications, not just for expectant mothers, but for all humans with impaired immune systems. They show that certain nutrients in the diet can strengthen one's immune system and resistance to disease. While this principle has been generally accepted for a long while, there are few studies specifically demonstrating how much certain vitamins and minerals can do to reverse or prevent disease. Many persons with diseases like cancer, AIDS, and CFS take vitamin supplements, since the treatments offered for these conditions are not yet proven and not always successful. In some cases, the supplements appear to help patients. Multiple vitamins are certainly a source of these nutrients, and certain of these vitamins and minerals are particularly helpful. Antioxidants Vitamins A, C, and E are helpful, along with Vitamins B6 and B12, calcium, chromium, iodine, iron, magnesium, selenium, and zinc.[6] It will be important to perform further studies to show how great the immune system benefits of supplements and vitamin and mineral-rich diets are, particularly in Americans hit hardest by radiation, such as the Baby Boomers.

Another body of evidence that demonstrates there is hope for deliverance from the low-level radiation scourge is contained in some of the statistics presented in this book. *When radiation levels in the environment decrease, health status improves.* Nowhere is this more evident than by the reduction in infant mortality and underweight births after the signing of the Partial Test Ban Treaty by Kennedy and Khrushchev in 1963. Levels of strontium-90 in the milk, for example, peaked at 30 picocuries per liter in May 1964 (see Chapter 3), and thereafter levels of this and other radioisotopes plunged steadily until about 1980, when radioactivity was less than 10% of what it had been in 1964. Statistics shown in Chapter 5 indicate what happened to the health of newborns during and after the time of Nevada above-ground bomb testing. From 1950 to 1966, the rate *fell just 23% and 13%*, respectively for whites and non-whites, while during the following 16 years (1966 to 1982), *plunges of 51% and 55%* occurred. This dramatic reduction induced Sternglass to estimate that an excess of 375,000 American infants had died because of atmospheric weapons testing. It is eerie to think about what might have happened to the infant mortality rate had atmospheric bomb tests continued. As it was, the no-progress years of the 1950s and early 1960s placed the U.S. well behind the rates of a number of other nations. We still have not recovered from that lost decade-and-a-half; today, the U.S. ranks behind at least 24 nations in infant mortality.[7]

Also in the post-bomb test period, the rate of births under 2500 grams stopped rising, and underwent a precipitous decline. From a 1950 to 1966 *rise of 2% and 35%* for whites and non-whites, respectively, the next 16 years saw *declines of 22% and 19%*. There were factors other than less dietary radiation aiding in this drop, such as the Medicaid program (enacted in 1966) and its greater access to adequate prenatal care for poor women. But again, reducing radiation is associated with a major improvement in the public's health. How much of the decline is due to other factors is impossible to quantify, but radiation reductions certainly are one of the reasons for these large improvements.

The closure of several nuclear power plants, and the cessation of all nuclear weapons production gives us another opportunity to examine how removing low-level radiation from the environment can improve the public's health. Weapons production was phased out from the late 1980s to the early 1990s, while power plants closed permanently at Rancho Seco in California (1989), Trojan in Oregon (1992), Yankee Rowe in Massachusetts (1993), Connecticut Yankee (1996), Maine Yankee (1996), and Big Rock Point in Michigan (1996). Others have been closed temporarily, such as the Millstone plant (not operating since March 1996). These dual trends, plus the cessation of underground atomic tests in Nevada in 1992, allow a preliminary analysis to be made of what, if any, effects this reduced radiation has made on human health. Again we will first address newborns, who are most sensitive to radiation and who often display changes in health status immediately after a change in radioactive emissions. The National Center for Health Statistics has estimated certain vital rates for 1996 in late 1997, enabling a meaningful analysis to be made.

The 1986 to 1991 drop of 13.2% in infant mortality was replaced by a greater drop of 19.1% from 1991 to 1996. Although this difference may not seem significant, the added drop in the latest five years saved about *5000 lives*. The states in which 1991 to 1996 declines in infant mortality were the greatest include:

- Massachusetts, location of the closed Yankee Rowe reactor, down 35%
- Oregon, location of the closed Trojan reactor, down 25%
- South Carolina, location of the closed Savannah River plant, down 31%
- Washington, location of the closed Hanford plant, down 31%

By contrast Tennessee fell only 11% during this time. While weapons production had ceased at Oak Ridge in the late 1980s, operations at a nearby civilian reactor and an onsite nuclear waste incinerator had begun (see Chapter 10), perhaps slowing progress. Data need to be looked at on a county-by-county basis, which is not readily available after 1992 as of early 1998. There are, however, examples in which lower emissions may be tied to fewer infant deaths. One is the trend in infant mortality near California's Rancho Seco reactor in its home county (Sacramento) and the four closest counties to its east (downwind). From 1985 to 1988, the infant mortality rate actually *rose* 14% in the five counties, while the U.S. rate dropped 6%. Then, after the reactor closed in 1989, an abrupt reversal took place. Between 1988 and 1992, the area's infant mortality rate *fell* 35%, compared to only a 15% reduction nationwide (Figure 11.1). This turnaround and excessive drop translates into 120 additional infants surviving the first year of life in the five counties.

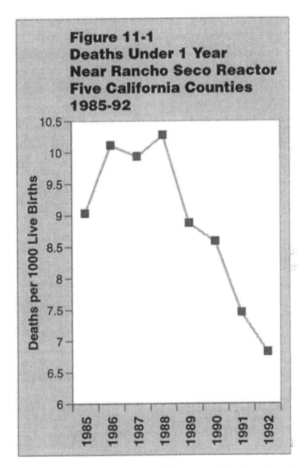

Figure 11-1
Deaths Under 1 Year
Near Rancho Seco Reactor
Five California Counties
1985-92

Another site already mentioned in Chapter 10 is the Fermi reactor near Detroit. After its relatively large releases in the late 1980s, emissions dropped and the plant was closed from December 1993 to April 1995. In Wayne County (home of Detroit), from 1992 to 1995, both the white and black infant mortality rates tumbled 35%, while the U.S. rate only went down 11%, a spectacular turnaround from the unchanging rates of the mid and late 1980s.

Reductions in underweight births have not yet taken place in states with closed plants. It will be interesting to perform trend analysis when county-specific data become available.

In summary, there has been a troubling set of health problems tied to radiation in the past. The problems have not gone away, and may continue well into the future. However, there is hope that recent changes have begun to reduce the danger, and that future changes will lower the threat even further. As Martin Luther King wrote from a Birmingham jail, "a tension of the mind" is critical to accomplishing this goal, even if the truth is upsetting and indicts the actions of many esteemed and well-meaning persons. The radiation health problem is a human one, and its solutions are human ones. Despite the arrogant and dangerous attitudes of leaders who brought us to the current mess, I have every confidence that as long as enough people remain committed to the truth, mankind will see the error of his ways and work to reduce the health threat to the millions who inhabit this nation.

References

1. *The New York Times*, January 2, 1998, D1.
2. *The New York Times*, August 2, 1997, 6.
3. Scholl, T. O., et al., Use of multivitamin/mineral prenatal supplements: influence on the outcome of pregnancy, *American Journal of Epidemiology*, 1997, 134-41.
4. Daly, S., et al., Minimum effective dose of folic acid for food fortification to prevent neural-tube defects, *Lancet*, December 6, 1997, 1666-9.
5. *The New York Times*, January 1, 1998, A12.
6. Shannon, S., *Diet for the Atomic Age*, Instant Improvement Inc., New York, 1993, 192.
7. National Center for Health Statistics, *Health United States 1996–97 and Injury Chartbook*, DHHS Publication (PHS) 97-1232, Hyattsville, MD, 1997, 105.

Index

Index

Printed and bound by CPI Group (UK) Ltd, Croydon, CR0 4YY

22/10/2024

01777605-0010